© 2016
Clement Ampadu
drampadu@hotmail.com

ISBN:978-1-365-43474-7
ID:19493838
www.lulu.com

All rights reserved. No part of this publication may be produced or transmitted in any form or by any means, electronic or mechanical, including photocopying and recording, or in any information storage and retrieval system, without the prior written permission of the publisher.

Contents

	Dedication	3
1	**Convergence of Common r-Fixed Point of Jungck Hardy Rogers Type Contractive Operators**	**4**
1.1	Brief Summary	4
1.2	Preliminaries	4
1.3	Main Results	8
1.3.1	r-th order Jungck-Hardy-Rogers-Type Contraction Mapping Theorem	8
1.3.2	Iteration of r-th order Jungck-Hardy-Rogers Type Contractions	10
1.4	Exercises	12
1.5	References	13
2	**A Generalization of Noor Iteration and a Stability Result for the Higher-Order Banach Contraction in Banach Spaces**	**14**
2.1	Brief Summary	14
2.2	Preliminaries	14
2.3	Main Results	16
2.4	Exercises	19
2.5	References	20
3	**Stability of the Higher-Order Hardy-Rogers Type Contraction in Multiplicative Cone Metric Space**	**21**
3.1	Brief Summary	21
3.2	Preliminaries	21
3.3	Main Results	23
3.4	Exercises	24
3.5	References	25
4	**Stability of the Higher-Order Jungck-Kannan Mapping Operator in b-Metric Space**	**26**
4.1	Brief Summary	26
4.2	Preliminaries	26
4.3	Main Results	29
4.4	Exercises	31
4.5	References	31
5	**Stability of Common r-Fixed Point Procedure for Higher-Order Generalizations of Banach, Kannan, and Chatterjea Contractions**	**33**
5.1	Brief Summary	33
5.2	Preliminaries	33
5.3	Main Results	36
5.4	Exercises	38
5.5	References	39

Dedication

This book is dedicated to those who read it .

Clement Ampadu
October, 2016

Chapter 1

Convergence of Common r-Fixed Point of Jungck Hardy Rogers Type Contractive Operators

1.1 Brief Summary

Abstract A.1 1

In the first part of this chapter we show the relationship between commuting maps and r-fixed points. Our work take inspiration from Proposition 1 [W. F. Pfeffer, More on involutions of a circle, this MONTHLY, 81 (1974) 613-616] which show the interdependence between commuting mappings and the notion of r-fixed points exist. We highlight this interdependence in a more general context, and obtain an r-fixed point in a similar way which will have the higher-order Hardy-Rogers-Type Mapping Theorem, Theorem A.2 [Ampadu, Clement (2016): Characterization Theorems Inspired by the Hardy-Rogers Map I: Some Results in Metric Spaces, lulu.com, ISBN:1365101185, 9781365101182] as a consequence. In the second part of this chapter we introduce higher-order Jungck type iteration procedure and a notion of higher-order weakly compatible mappings. We then obtain a general stability result for what we call common higher-order fixed points by an implicit contraction condition satisfying a so-called (r-E. A) property. A consequence of this general stability result is a stability result for the higher-order Jungck-Hardy-Rogers type mapping theorem

1.2 Preliminaries

Definition A.1 1

Let (X, d) be a metric space. The pair (g, f) of self-maps on X will be said to form a higher-order Jungck-Hardy-Rogers type contraction if it holds that $d(g^r x, g^r y) \leq Z^\star \beta^r [d(fx, gx) + d(fy, gy) + d(fx, gy) + d(fy, gx) + d(fx, fy)]$ for all $x, y \in X$ where $\beta \in [0, \frac{1}{5})$ but $Z^\star$ is a certain modification on Proposition A.3 [Ampadu, Clement (2016): Characterization Theorems Inspired by the Hardy-Rogers Map I: Some Results in Metric Spaces, lulu.com, ISBN:1365101185, 9781365101182]

Remark A.2 1

If f is the identity mapping in the above definition then we recover Definition A.5 [Ampadu, Clement (2016): Characterization Theorems Inspired by the Hardy-Rogers Map I: Some Results in Metric Spaces, lulu.com, ISBN:1365101185, 9781365101182], that is, higher-order Jungck-Hardy-Rogers type contraction becomes higher-order Hardy-Rogers type contraction

Remark A.3 1

If $y = g^r x$ in the above definition, then employing similar technique used in obtaining Lemma A.1 [Ampadu, Clement (2016): Characterization Theorems Inspired by the Hardy-Rogers Map I: Some Results in Metric Spaces, lulu.com, ISBN:1365101185, 9781365101182], we deduce that the higher-order Jungck-Hardy-Rogers type contraction can be written as $d(g^r x, g^{2r} x) \leq \gamma d(fx, g^r x)$, where $\gamma := \frac{3Z^* \beta^r}{1 - 2Z^* \beta^r}$

Since $\gamma < 1$, it follows immediately from the remark above, that, we have the following

Lemma A.4 1

Let $\{y_{rn}\}$ be a sequence in a complete metric space (X, d). If there exists $\gamma < 1$ such that $d(y_{r(n+1)}, y_{rn}) \leq \gamma d(y_{rn}, y_{r(n-1)})$ for all n and any $r \in \mathbb{N}$, then $\{y_{rn}\}$ converges to a point in X

Definition A.5 1

Let (X, d) be a metric space and $S, T : X \mapsto X$ be two mappings. We say that S and T are r-commuting if $S^r T^r x = T^r S^r x$ for all $x \in X$ and any $r \in \mathbb{N}$

Taking inspiration from [S. Sessa, "On a weak commutativity condition of mappings in fixed point considerations," Publ. Inst. Math., Nouv. Ser., vol. 32(46), pp. 149–153, 1982] we introduce a generalization of r-commuting mappings as follows

Definition A.6 1

Let (X, d) be a metric space and $S, T : X \mapsto X$ be two mappings. We say that S and T are weakly r-commuting if $d(S^r T^r x, T^r S^r x) \leq d(S^r x, T^r x)$ for all $x \in X$ and any $r \in \mathbb{N}$

Taking inspiration from [G. Jungck, "Compatible mapping and common fixed points," Int. J. Math. Math. Sci., vol. 9, pp. 771–779, 1986] we introduce a generalization of weakly r-commuting maps as follows

Definition A.7 1

Let (X, d) be a metric space and $S, T : X \mapsto X$ be two mappings. We say S and T are r-compatible if $d(S^r T^r x_n, T^r S^r x_n) \to 0$ as $n \to \infty$ whenever $\{x_n\}$ is a sequence in X such that $S^r x_n \to t$ and $T^r x_n \to t$ as $n \to \infty$, for some $t \in X$

Remark A.8 1

r-commuting implies weakly r-commuting which in turn implies r-compatibility, but the converse is not true in general

Example A.9 1

Let $X = \mathbb{R}$ and define the functions $f, g : X \mapsto X$ by $f^r(x) = x$ and $g^r(x) = 2^r x$ for any $r \in \mathbb{N}$. Notice that $|f^r(g^r x_n) - g^r(f^r x_n)| = 0$ for any sequence $\{x_n\}$ in X, but $|f^r(x_n) - g^r(x_n)| = |1 - 2^r||x_n|$, therefore if $\{x_n\}$ is a sequence converging to zero, then f and g are r-compatible. Also notice that f and g are weakly r-commuting maps, since, $|f^r(g^r x) - g^r(f^r x)| = 0 \leq |1 - 2^r||x| = |f^r x - g^r x|$ for all $x \in X$ and any $r \in \mathbb{N}$

Taking inspiration from [M. Aamri and D. El Moutawakil, "Some new common fixed point theorems under strict contractive conditions," J. Math. Anal. Appl., vol. 270, no. 1, pp. 181–188, 2002] we introduce a notion which is independent of the notion of weakly r-compatibility

> **Definition A.10 1**
>
> Let (X,d) be a metric space and $S,T : X \mapsto X$ be two mappings. We say they satisfy the $(r$-E.A$)$ property if there exists a sequence $\{x_n\}$ in X such that $S^r x_n \to t$ and $T^r x_n \to t$ as $n \to \infty$ for some $t \in X$ and any $r \in \mathbb{N}$

> **Example A.11 1**
>
> Consider $(\mathbb{R}_+, |\cdot|)$ and define S and T by $S^r x = 2^r x$ and $T^r x = x$ for any $r \in \mathbb{N}$. Notice that $S^r x = T^r x$ iff $x = 0$. Let $\{x_n\}$ be a sequence in X given by $x_n = \frac{1}{n}$, $n \geq 1$. Notice that $S^r x_n \to 0$ and $T^r x_n \to 0$ as $n \to \infty$. Thus, S and T satisfy $(r$-E.A$)$ property. Moreover, S and T are weakly r-compatible since $S^r(T^r 0) = T^r(S^r 0) = 0$

> **Example A.12 1**
>
> Consider $X = [-1, 1]$ with the usual metric. Define $S, T : X \mapsto X$ for any $r \in \mathbb{N}$ as follows: $T^r(x) = \frac{1}{2}$ if $x = -1$; $T^r(x) = \frac{x}{4^r}$ if $x \in (-1, 1)$; $T^r(x) = \frac{3}{5}$ if $x = 1$; $S^r(x) = \frac{1}{2}$ if $x = -1$; $S^r(x) = \frac{x}{2^r}$ if $x \in (-1, 1)$; $S^r(x) = \frac{-1}{2}$ if $x = 1$. Let $\{x_n\}$ be given by $x_n = \frac{1}{n}$, then, $S^r x_n \to 0$ and $T^r x_n \to 0$ as $n \to \infty$. It follows that the pair (S,T) satisfy the $(r$-E.A$)$ property. On the other hand, notice that $T^r(X) = \{\frac{1}{2}, \frac{3}{5}\} \cup (\frac{-1}{4^r}, \frac{1}{4^r})$ and $S^r(X) = \{\frac{-1}{2}, \frac{1}{2}\} \cup (\frac{-1}{2^r}, \frac{1}{2^r})$. Hence, neither $T^r(X)$ is contained in $S^r(X)$ nor $S^r(X)$ is contained in $T^r(X)$. This example shows that a pair satisfying $(r$-E.A$)$ property need not follow the pattern of containment of range of one map into the range of another as is utilized in common r-fixed point considerations but still relaxes such requirements

Taking inspiration from [G. Jungck, "Commuting mappings and fixed points," Am. Math. Mon., vol. 83, pp. 261–263, 1976] we introduce the following

> **Definition A.13 1**
>
> Let S and T be mappings from X into itself, and $T^r(X) \subseteq S^r(X)$ for any $r \in \mathbb{N}$. Fix $x_0 \in X$, the iterative procedure $S^r x_{n+1} = T^r x_n$, for any $r \in \mathbb{N}$ and $n = 0, 1, 2, \cdots$ will be called the higher-order Jungck iterative procedure

Taking inspiration from [S. L. Singh and B. Prasad, "Some coincidence theorems and stability of iterative procedures," Comput. Math. Appl., vol. 55, no. 11, pp. 2512–2520, 2008] we introduce the following

> **Definition A.14 1**
>
> Let (X, d) be a metric space and $S, T : X \mapsto X$. Let z be a r-coincidence point of T and S, that is, $S^r z = T^r z = t$. Fix $x_0 \in X$ and let $\{S^r x_n\}$ generated by the higher-order Jungck iterative procedure $S^r x_{n+1} = T^r x_n$, for any $r \in \mathbb{N}$ and $n = 0, 1, 2, \cdots$ converge to $t \in X$. Let $\{S^r y_n\} \subset X$ be an arbitrary sequence and set $\epsilon_n = d(S^r y_{n+1}, T^r y_n)$, $n = 0, 1, 2, \cdots$. Then the higher-order Jungck iterative procedure will be called r-(S,T)-stable or r-stable with respect to (S,T) iff $\epsilon_n \to 0$ as $n \to \infty$ implies $S^r y_n \to t$ as $n \to \infty$ for any $r \in \mathbb{N}$

In the sequel we will need the following result contained in [V. Berinde, Iterative approximation of fixed points. 2nd revised and enlarged ed., ser. Lecture Notes in Mathematics. Berlin: Springer, 2007, vol. 1912]

> **Lemma A.15 1**
>
> Let $\{a_n\}_{n=0}^{\infty}$ and $\{b_n\}_{n=0}^{\infty}$ be sequences of nonnegative numbers and let $0 \leq h < 1$ be a constant. Suppose $a_{n+1} \leq h a_n + b_n$, $n \geq 0$
>
> (a) If $b_n \to 0$ as $n \to \infty$, then, $a_n \to 0$ as $n \to \infty$
>
> (b) If $\sum_{n=0}^{+\infty} b_n < \infty$, then, $\sum_{n=0}^{+\infty} a_n < \infty$

Popa in a series of papers [V. Popa, "Fixed point theorems for implicit contractive mappings," Stud. Cercet. Stiint ., Ser. Mat., Univ. Bacau, vol. 7, pp. 127–133, 1997; V. Popa, "Some fixed point theorems for compatible mappings satisfying an implicit relation," Demonstr. Math., vol. 32, no. 1, pp. 157–163, 1999; V. Popa, "A generalized fixed point theorem for two pairs of mappings on two metric spaces," Novi Sad J. Math., vol. 35, no. 2, pp. 79–83, 2005] introduced a class of implicit functions which allows one to write contractive type conditions in terms of them. We recall the class of implicit functions due to Popa.

Definition A.16 1

A function $F : \mathbb{R}_+^5 \mapsto \mathbb{R}$ will be called a **Popa class function** if it satisfies the following conditions:

(1) F is continuous in each coordinate variable

(2) If for some $u, v, w \geq 0$, we have

   (a) $F(u, v, u, v, w) \leq 0$

   (b) $F(u, v, v, u, w) \leq 0$

then there exists $h \in [0, 1)$, such that $u \leq h \max\{v, w\}$

(3) $F(u, u, u, u, 0) > 0$, for all $u > 0$

Remark A.17 1

We will denote by P_c the class of all functions satisfying the above definition. Some elements of P_c are contained in [V. Popa, "A generalized fixed point theorem for two pairs of mappings on two metric spaces," Novi Sad J. Math., vol. 35, no. 2, pp. 79–83, 2005; H. K. Pathak and R. K. Verma, "Coincidence and common fixed points in symmetric spaces under implicit relation and application," Int. Math. Forum, vol. 3, no. 29-32, pp. 1489–1499, 2008]

The following examples will be useful in the sequel

Example A.18 1

The function $F(t_1, t_2, t_3, t_4, t_5) : \mathbb{R}_+^5 \mapsto \mathbb{R}$ given by $F(t_1, \cdots, t_5) = t_1 - ft_2 - \frac{c+e}{2}t_3 - \frac{c+e}{2}t_4 - (a+b)t_5$, where $a, b, c, e, f \in [0, 1)$, with $a + b + c + e + f < 1$ satisfies

(i) (1) and (2a) of Definition A.16 with $h = \frac{c+e+2f}{2-c-e} \in [0, 1)$

(ii) (2b) of Definition A.16 with $h = \frac{c+e+2f}{2-c-e} \in [0, 1)$

(iii) (3) of Definition A.16, where $h = \frac{c+e+2f}{2-c-e} \in [0, 1)$, if $\max\{v, w\} = v$ and $h = \frac{c+e+2f}{2-c-e} \in [0, 1)$, if $\max\{v, w\} = w$

Example A.19 1

The function $F(t_1, t_2, t_3, t_4, t_5) : \mathbb{R}_+^5 \mapsto \mathbb{R}$ given by $F(t_1, \cdots, t_5) = t_1 - y[t_2 + t_3 + t_4 + 2t_5]$, where $y \in [0, \frac{1}{5})$, satisfies

(i) (1) and (2a) of Definition A.16 with $h = \frac{4y}{2-2y} \in [0, \frac{1}{5})$

(ii) (2b) of Definition A.16 with $h = \frac{4y}{2-2y} \in [0, \frac{1}{5})$

(iii) (3) of Definition A.16, where $h = \frac{4y}{2-2y} \in [0, \frac{1}{5})$, if $\max\{v, w\} = v$ and $h = \frac{4y}{2-2y} \in [0, \frac{1}{5})$, if $\max\{v, w\} = w$

1.3 Main Results

1.3.1 r-th order Jungck-Hardy-Rogers-Type Contraction Mapping Theorem

First we begin with the following

Proposition A.1 1

Let f be a mapping of a set X into itself. Then f has a r-fixed point iff there is a constant map $h : X \mapsto X$ which r-commutes with f, that is, $h^r(f^r(x)) = f^r(h^r(x))$ for all $x \in X$ and any $r \in \mathbb{N}$

Proof of Proposition A.1 1

(Sufficiency) By hypothesis there exists $a \in X$ and $h : X \mapsto X$ such that $h^r(x) = a$ and $h^r(f^r(x)) = f^r(h^r(x))$ for any $r \in \mathbb{N}$ and all $x \in X$. It follows that we can write $f^r(a) = f^r(h^r(a)) = h^r(f^r(a)) = a$, so that a is a r-fixed point of f

(Necessity) Suppose $f^r(a) = a$ for some $a \in X$ and any $r \in \mathbb{N}$. Define $g : X \mapsto X$ by $g^r(x) = a$ for all $x \in X$ and any $r \in \mathbb{N}$. Then, $g^r(f^r(x)) = a$ and $f^r(g^r(x)) = a$ for all $x \in X$ and any $r \in \mathbb{N}$, it follows that, $g^r(f^r(x)) = f^r(g^r(x))$ for all $x \in X$ and any $r \in \mathbb{N}$, and thus g and f are r-commuting

Theorem A.2 1

Let f be a r-continuous mapping of a complete metric space (X, d) into itself, that is, $f^r : X \mapsto X$ is continuous for any $r \in \mathbb{N}$. Then f has an r-fixed point in X iff it holds that $d(g^r x, g^{2r} x) \leq Z^{**} \beta^r [d(f^r x, gx) + d(f^r g^r x, g^{r+1} x) + d(f^r x, g^{r+1} x) + d(f^r g^r x, gx) + d(f^r x, f^r g^r x)]$ for all $x \in X$ where $\beta \in [0, \frac{1}{5})$ but Z^{**} is a certain modification on Proposition A.3 [Ampadu, Clement (2016): Characterization Theorems Inspired by the Hardy-Rogers Map I: Some Results in Metric Spaces, lulu.com, ISBN:1365101185, 9781365101182], $g : X \mapsto X$ r-commutes with f and $g^r(X) \subset f^r(X)$. Moreover, if $g^r(X) \subset f^r(X)$ and the inequality holds for all $x \in X$, then f and g have a unique common r-fixed point.

Proof of Theorem A.2 1

The necessity portion of the proof follows from Proposition A.2. Moreover, $g^r(x) = a = f^r(a)$ for all $x \in X$, thus, $g^r(X) \subset f^r(X)$. By Remark A.3, the contractive inequality of the theorem can be written as $d(g^r x, g^{2r} x) \leq \gamma d(f^r x, g^r x)$, where $\gamma := \frac{3Z^{**}\beta^r}{1-2Z^{**}\beta^r}$. Since $0 \leq \gamma < 1$, we have for all $x \in X$ that

$$d(g^r x, g^{2r} x) = d(a, a) = 0 \leq \gamma d(f^r x, g^r x)$$

thus, $g^r(X) \subset g^r(X)$ and the contractive inequality holds.

Now suppose that there is a mapping g of X into itself which r-commutes with f and for which $g^r(X) \subset g^r(X)$ and the contractive inequality holds. We now show that this condition is sufficient to ensure that f and g have a unique common r-fixed point. Now let $x_0 \in X$ and let x_1 be such that $g^r(x_0) = f^r x_1$. In general, choose x_n so that $f^r x_n = g^r x_{n-1}$. We can do this since $g^r(X) \subset f^r(X)$. Now one deduces that

$$\begin{aligned} d(f^r x_{n+1}, f^r x_n) &= d(g^r x_n, g^r x_{n-1}) \\ &= d(g^{2r} x_{n-1}, g^r x_{n-1}) \\ &\leq \gamma d(f^r x_{n-1}, g^r x_{n-1}) \\ &= \gamma d(f^r x_{n-1}, f^r x_n) \end{aligned}$$

By Lemma A.4, there is $t \in X$ such that $f^r x_n \to t$, but since, $f^r x_n = g^r x_{n-1}$, it follows also that $g^r x_n \to t$. Now since f is r-continuous, the contractive inequality of the theorem implies both f and g are r-continuous. It follows that $g^r(f^r x_n) \to g^r t$ and $f^r(g^r x_n) \to f^r t$. But f and g r-commute, thus, $f^r(g x_n) = g^r(f x_n)$ for all n, thus, in the limit, $f^r t = g^r t$. Consequently, $f^{2r} t = f^r(g^r t) = g^{2r} t$ by r-commutativity. It follows that $d(g^r t, g^{2r} t) \leq \gamma d(f^r t, f^r(g^r t)) = \gamma d(g^r t, g^{2r} t)$. Since $1 - \gamma \neq 0$, it follows that, $d(g^r t, g^{2r} t) = 0$, that is, $g^r t = g^{2r} t$. We now have, $g^r t = g^{2r} t = f^r(g^r t)$, that is, $g^r t$ is the common r-fixed point of f and g. Finally we show uniqueness of the common r-fixed point. Suppose that $x = f^r x = g^r x$ and $g^r x = g^{2r} x = f^r(g^r x)$ but $x \neq g^r x$, then, $d(x, g^r x) = d(g^r x, g^{2r} x) \leq \gamma d(f^r x, f^r(g^r x)) = \gamma d(x, g^r x)$, since $1 - \gamma \neq 0$, we get that, $d(x, g^r x) = 0$ which implies that $x = g^r x$, a contradiction. Thus, uniqueness of the common r-fixed point follows.

Corollary A.3 1

Let f and g be r-commuting maps of a complete metric space (X, d) into itself. Suppose that f is r-continuous and $g^r(X) \subset f^r(X)$. Suppose there exists a positive integer k and any $r \in \mathbb{N}$ such that $d(g^{kr} x, g^{2kr} x) \leq Z^{**} \beta^r [d(f^r x, g x) + d(f^r g^r x, g^{r+1} x) + d(f^r x, g^{r+1} x) + d(f^r g^r x, g x) + d(f^r x, f^r g^r x)]$ for all $x \in X$ where $\beta \in [0, \frac{1}{5})$ but Z^{**} is a certain modification on Proposition A.3 [Ampadu, Clement (2016): Characterization Theorems Inspired by the Hardy-Rogers Map I: Some Results in Metric Spaces, lulu.com, ISBN:1365101185, 9781365101182], then f and g have a unique common r-fixed point.

Proof of Corollary A.3 1

Proof. Its clear that g^k r-commutes with f and $g^{rk}(X) \subset g^r(X) \subset f^r(X)$. Thus applying the previous theorem to g^k and f it follows that there exists a unique $a \in X$ such that $a = f^r a = g^{rk} a$. However, since f and g r-commute, we can write $g^r(a) = f^r(g^r a) = g^{rk}(g^r(a))$, thus, $g^r(a)$ is a common r-fixed point of f and g^k. The uniqueness of a implies that $a = g^r a = f^r a$ □

> **Remark A.4 1**
>
> Due to Remark A.2 and Remark A.3, if take $k=1$ and f the identity mapping in the previous Corollary, then we obtain the higher-order Hardy-Rogers Type Mapping Theorem, Theorem A.2 [Ampadu, Clement (2016): Characterization Theorems Inspired by the Hardy-Rogers Map I: Some Results in Metric Spaces, lulu.com, ISBN:1365101185, 9781365101182] as a consequence.

1.3.2 Iteration of r-th order Jungck-Hardy-Rogers Type Contractions

> **Theorem A.5 1**
>
> Let (X,d) be a complete metric space and $S,T: X \mapsto X$ be two mappings, such that S,T satisfy $(r\text{-E.A})$ property and $S^r(X)$, for any $r \in \mathbb{N}$, is a complete subspace of X. Suppose there exists $F \in P_c$ such that
>
> $$F(d(T^r x, T^r y), d(Sx, Sy), d(Sx, Ty), d(Sy, Tx), \frac{d(Sx, Tx) + d(Sy, Ty)}{2}) \leq 0$$
>
> for all $x, y \in X$ and any $r \in \mathbb{N}$. If
>
> (i) F satisfies (2b) of Definition A.16, then the pair (S,T) have a point of r-coincidence
>
> (ii) F satisfies (3) of Definition A.16, then the pair (S,T) have a common r-fixed point provided they are weakly r-compatible
>
> (iii) F satisfies (3) of Definition A.16, then the associated iterative procedure is r-stable with respect to (S,T)

CHAPTER 1. CONVERGENCE OF COMMON R-FIXED POINT OF JUNGCK HARDY ROGERS TYPE CONTRACTIVE OPERATORS

> **Proof of Theorem A.5 1**
>
> Since T and S satisfy (r-E.A) property, there exists a sequence $\{x_n\}$ in X such that $T^r x_n \to t$ and $S^r x_n \to t$ as $n \to \infty$ for some $t \in X$. Since $S^r(X)$ is a complete subspace of X for any $r \in \mathbb{N}$, every convergent sequence of points of $S^r(X)$ has a limit in $S^r(X)$ for any $r \in \mathbb{N}$. Therefore, $S^r x_n \to t = S^r z$ and $T^r x_n \to t = S^r z$, which in turn implies that $t = S^r z \in S^r(X)$ for any $r \in \mathbb{N}$. Now we claim that $S^r z = T^r z$ for any $r \in \mathbb{N}$. If not, then, $d(S^r z, T^r z) > 0$. Now by the implicit contractive condition of the theorem, we have,
>
> $$F(d(T^r z, T^r x_n), d(Sz, Sx_n), d(Sz, Tx_n), d(Sx_n, Tz), \frac{d(Sz, Tz) + d(Sx_n, Tx_n)}{2}) \leq 0$$
>
> but,
>
> $$F(d(T^r z, T^r x_n), d(Sz, Sx_n), d(Sz, Tx_n), d(Sx_n, Tz), \frac{d(Sz, Tz) + d(Sx_n, Tx_n)}{2}) \leq$$
> $$F(d(T^r z, T^r x_n), d(S^r z, S^r x_n), d(S^r z, T^r x_n), d(S^r x_n, T^r z), \frac{d(S^r z, T^r z) + d(S^r x_n, T^r x_n)}{2})$$
>
> From the two inequalities immediately above, one deduces that,
>
> $$F(d(T^r z, T^r x_n), d(S^r z, S^r x_n), d(S^r z, T^r x_n), d(S^r x_n, T^r z), \frac{d(S^r z, T^r z) + d(S^r x_n, T^r x_n)}{2}) \leq 0$$
>
> If we take limits in the above inequality as $n \to \infty$, we deduce that,
>
> $$F(d(T^r z, S^r z), 0, 0, d(S^r z, T^r z), \frac{d(S^r z, T^r z)}{2}) \leq 0$$
>
> According to (2b) of Definition A.16, there exists $h \in [0, 1)$ such that
>
> $$d(T^r z, S^r z) \leq h \max\{0, \frac{d(S^r z, T^r z)}{2}\} = h \frac{d(S^r z, T^r z)}{2} < d(T^r z, S^r z)$$
>
> which is a contradiction. Thus, $S^r z = T^r z$, so z is a r-coincidence point of T and S. Since S and T are weakly r-compatible, then, $S^r t = S^r T^r z = T^r S^r z = T^r t$. Now we claim that $T^r t = t$. If not, then, $d(T^r t, t) > 0$, and using the implicit contractive condition of the theorem, we deduce that $F(d(T^r t, t), d(T^r t, t), d(T^r t, t), d(T^r t, t), 0) \leq 0$, which contradicts (3) of Definition A.16. Thus, $T^r t = t$, which shows that t is a common r-fixed point of T and S. The uniqueness of the common r-fixed follows from the implicit contractive condition of the theorem. Now we prove r-(S,T)-stability. Let $\{S^r y_n\} \subset X$ be an arbitrary sequence and set $\epsilon_n = d(S^r y_{n+1}, T^r y_n)$. Assume that $\epsilon_n \to 0$ as $n \to \infty$. By the triangle inequality, we have, $d(S^r y_{n+1}, t) \leq d(S^r y_{n+1}, T^r y_n) + d(T^r y_n, t) = \epsilon_n + d(T^r y_n, t)$. With $x := t$ and $y := y_n$ in the implicit contractive condition of the theorem, we deduce that
>
> $$F(d(T^r y_n, t), d(S^r y_n, t), d(T^r y_n, t), d(S^r y_n, t), \frac{d(S^r y_n, T^r y_n)}{2}) \leq 0$$
>
> Since F satisfies (2a) of Definition A.16, it follows that there exists $h \in [0, 1)$ such that $d(T^r y_n, t) \leq h \max\{d(S^r y_n, t), \frac{d(S^r y_n, T^r y_n)}{2}\}$. If $\max\{d(S^r y_n, t), \frac{d(S^r y_n, T^r y_n)}{2}\} = d(S^r y_n, t)$, then, $d(T^r y_n, t) \leq h d(S^r y_n, t)$. An application of the triangle inequality implies that $d(S^r y_{n+1}, t) \leq h d(S^r y_n, t) + \epsilon_n$, and from Lemma A.15, it follows that $S^r y_n \to t$ as $n \to \infty$. If $\max\{d(S^r y_n, t), \frac{d(S^r y_n, T^r y_n)}{2}\} = \frac{d(S^r y_n, T^r y_n)}{2}$, then an application of the triangle inequality implies that $d(T^r y_n, t) \leq \frac{h}{2}[d(T^r y_n, t) + d(t, S^r y_n)]$, from which one deduces that $d(T^r y_n, t) \leq q d(S^r y_n, t)$, where $q := \frac{\frac{h}{2}}{1 - \frac{h}{2}}$. An application of the triangle inequality implies that $d(S^r y_{n+1}, t) \leq q d(S^r y_n, t) + \epsilon_n$, and from Lemma A.15, it follows that $S^r y_n \to t$ as $n \to \infty$.

Now we obtain a stability result for the higher-order Jungck-Hardy-Rogers type Mapping Theorem as follows

Corollary A.6 1

Let (X,d) be a complete metric space and $S, T : X \mapsto X$ be two mappings, such that T and S satisfy (r-E.A) property and $S^r(X)$ is a complete subspace of X for any $r \in \mathbb{N}$. Suppose there exists $F \in P_c$ such that F satisfies the implicit contractive condition of the previous theorem, for all $x, y \in X$ and any $r \in \mathbb{N}$. Then in the case of the higher-order Hardy-Rogers type contraction conditions, the associated common r-fixed point iterative procedure is r-stable with respect to (S,T).

Proof of Corollary A.6 1

Let F be given by Example A.19 and set $y := Z\beta^r$ for any $r \in \mathbb{N}$, where $\beta \in [0, \frac{1}{5})$ and $Z \geq 1$ is a certain modification on the bound in Proposition A.3 [Ampadu, Clement (2016): Characterization Theorems Inspired by the Hardy-Rogers Map I: Some Results in Metric Spaces, lulu.com, ISBN:1365101185, 9781365101182]. By applying the previous theorem we get the stability result for the higher-order Jungck-Hardy-Rogers type mapping theorem given by Theorem A.2 above or equivalently a stability result for the higher-order Hardy-Rogers type mapping theorem, Theorem A.2 [Ampadu, Clement (2016): Characterization Theorems Inspired by the Hardy-Rogers Map I: Some Results in Metric Spaces, lulu.com, ISBN:1365101185, 9781365101182] corresponding to a pair of mappings with a common r-fixed point

1.4 Exercises

Exercise A.1 1

Prove that the following is a particular case of Theorem 2 [Ioana Timis, STABILITY OF JUNGCK-TYPE ITERATIVE PROCEDURE FOR SOME CONTRACTIVE TYPE MAPPINGS VIA IMPLICIT RELATIONS, Miskolc Mathematical Notes Vol. 13 (2012), No. 2, pp. 555–567]: If F is given by Example A.18, then we obtain a stability result for the Hardy and Rogers fixed point theorem [G. E. Hardy and T. D. Rogers, "A generalization of a fixed point theorem of Reich," Can. Math. Bull., vol. 16, pp. 201–206, 1973] corresponding to a pair of mappings with a common fixed point

Exercise A.2 1

Prove that the following is a particular case of Theorem A.5: If F is given by Example 4 [Ioana Timis, STABILITY OF JUNGCK-TYPE ITERATIVE PROCEDURE FOR SOME CONTRACTIVE TYPE MAPPINGS VIA IMPLICIT RELATIONS, Miskolc Mathematical Notes Vol. 13 (2012), No. 2, pp. 555–567; see also, V. Popa, "A generalized fixed point theorem for two pairs of mappings on two metric spaces," Novi Sad J. Math., vol. 35, no. 2, pp. 79–83, 2005] with $a := M\lambda^r$ for any $r \in \mathbb{N}$, where $M \geq 1$ is the bound given by Proposition 4.1 contained in [Ezearn Fixed Point Theory and Applications (2015) 2015:88] and $\lambda \in [0,1)$, then we obtain a stability result for the higher-order Banach Contraction Mapping Theorem [Ampadu, Clement (2015): Generalization of Higher Order Contraction Mapping Theorem. Unpublished] corresponding to a pair of mappings with a common fixed point.

Exercise A.3 1

Show that the higher-order Banach Contraction Mapping Theorem [Ampadu, Clement (2015): Generalization of Higher Order Contraction Mapping Theorem. Unpublished] corresponding to a pair of mappings with a common fixed point is the higher-order version of the Jungck Contraction Principle [Gerald Jungck, Commuting Mappings and Fixed Points, The American Mathematical Monthly, Vol. 83, No. 4 (Apr., 1976), pp. 261-263]

1.5 References

(1) W. F. Pfeffer, More on involutions of a circle, this MONTHLY, 81 (1974) 613-616

(2) Ampadu, Clement (2016): Characterization Theorems Inspired by the Hardy-Rogers Map I: Some Results in Metric Spaces, lulu.com, ISBN:1365101185, 9781365101182

(3) S. Sessa, "On a weak commutativity condition of mappings in fixed point considerations," Publ. Inst. Math., Nouv. Ser., vol. 32(46), pp. 149–153, 1982

(4) G. Jungck, "Compatible mapping and common fixed points," Int. J. Math. Math. Sci., vol. 9, pp. 771–779, 1986

(5) M. Aamri and D. El Moutawakil, "Some new common fixed point theorems under strict contractive conditions," J. Math. Anal. Appl., vol. 270, no.1, pp. 181–188, 2002

(6) G. Jungck, "Commuting mappings and fixed points," Am. Math. Mon., vol. 83, pp. 261–263, 1976

(7) S. L. Singh and B. Prasad, "Some coincidence theorems and stability of iterative procedures," Comput. Math. Appl., vol. 55, no. 11, pp. 2512–2520, 2008

(8) V. Berinde, Iterative approximation of fixed points. 2nd revised and enlarged ed., ser. Lecture Notes in Mathematics. Berlin: Springer, 2007, vol. 1912

(9) V. Popa, "Fixed point theorems for implicit contractive mappings," Stud. Cercet. Stiint., Ser. Mat., Univ. Bacau, vol. 7, pp. 127–133, 1997

(10) V. Popa, "Some fixed point theorems for compatible mappings satisfying an implicit relation,"Demonstr. Math., vol. 32, no. 1, pp. 157-163, 1999

(11) V. Popa, "A generalized fixed point theorem for two pairs of mappings on two metric spaces," Novi Sad J. Math., vol. 35, no. 2, pp. 79–83, 2005

(12) H. K. Pathak and R. K. Verma, "Coincidence and common fixed points in symmetric spaces under implicit relation and application," Int. Math. Forum, vol. 3, no. 29-32, pp. 1489–1499, 2008

(13) Ioana Timis, STABILITY OF JUNGCK-TYPE ITERATIVE PROCEDURE FOR SOME CONTRACTIVE TYPE MAPPINGS VIA IMPLICIT RELATIONS, Miskolc Mathematical Notes Vol. 13 (2012), No. 2, pp. 555–567

(14) G. E. Hardy and T. D. Rogers, "A generalization of a fixed point theorem of Reich," Can. Math. Bull., vol. 16, pp. 201–206, 1973

(15) Ezearn Fixed Point Theory and Applications (2015) 2015:88

(16) Ampadu, Clement (2015): Generalization of Higher Order Contraction Mapping Theorem. Unpublished

(17) Gerald Jungck, Commuting Mappings and Fixed Points, The American Mathematical Monthly, Vol. 83, No. 4 (Apr., 1976), pp. 261-263

Chapter 2

A Generalization of Noor Iteration and a Stability Result for the Higher-Order Banach Contraction in Banach Spaces

2.1 Brief Summary

> **Abstract B.1 1**
>
> The Noor iteration was introduced in [Noor, M. A.: New approximations schemes for general variational inequalities. J. Math. Anal. Appl. 251 (2000), 217–299], and the Ishikawa iteration was introduced in [S. ISHIKAWA, Fixed points by a new iteration method, Proc. Amer. Math. Soc. 44 (1974),147-150]. In this chapter, generalizations of the Noor and Ishikawa iterations are introduced, and we prove stability for the generalization of Noor iteration for a class of functions satisfying a higher-order Banach contractive type condition. As a special case, we also establish a similar result for the generalization of the Ishikawa iteration.

2.2 Preliminaries

> **Notation B.1 1**
>
> Let (E, d) be a complete metric space, and $T : E \mapsto E$. The set of r-fixed points of T in E for any $r \in \mathbb{N}$ will be denoted by $F_{T^r} = \{p \in E : T^r p = p\}$

> **Definition B.2 1**
>
> Let $\{x_n\}_{n=0}^{\infty} \subset E$ be the sequence generated by an iteration procedure involving the operator T, that is, $x_{n+1} = f(T^r, x_n)$, for all $n = 0, 1, \cdots$, and any $r \in \mathbb{N}$, where $x_0 \in E$ is the initial approximation and f is some function. Suppose $\{x_n\}_{n=0}^{\infty} \subset E$ converges to a r-fixed point p of T in E. Let $\{y_n\}_{n=0}^{\infty} \subset E$ and set $\epsilon_n = d(y_{n+1}, f(T^r, y_n))$, for all $n = 0, 1, 2, \cdots$ and any $r \in \mathbb{N}$, then we say that $x_{n+1} = f(T^r, x_n)$, for all $n = 0, 1, \cdots$, and any $r \in \mathbb{N}$, is r-T-stable or r-stable with respect to T if $\epsilon_n \to 0$ as $n \to \infty$ implies $y_n \to p$ as $n \to \infty$

> **Remark B.3 1**
>
> Since metric is induced by norm, we can write, $\epsilon_n = \|y_{n+1} - f(T^r, y_n)\|$ for all $n = 0, 1, 2, \cdots$ and any $r \in \mathbb{N}$, whenever E is a normed linear space or a Banach space

Remark B.4 1

If $f(T^r, x_n)$ is given by $f(T^r, x_n) := T^r x_n$ for all $n = 0, 1, 2, \cdots$ and any $r \in \mathbb{N}$, then we get a generalization of the Picard iteration process, which we will call the higher-order Picard iteration process. If $r = 1$ in the higher-order Picard iteration process, then we get the original Picard iteration process

Remark B.5 1

If $f(T^r, x_n)$ is given by $f(T^r, x_n) := (1 - \alpha_n)x_n + \alpha_n T^r x_n$ with $\{\alpha_n\}_{n=0}^{\infty}$ a sequence of real numbers in $[0, 1]$, then we get a generalization of the Mann iteration process, which we will call the higher-order Mann iteration process. If $r = 1$ in the higher-order Mann iteration process, then we get the original Mann iteration process

Definition B.6 1

Let $\{\alpha_n\}_{n=0}^{\infty}$ and $\{\beta_n\}_{n=0}^{\infty}$ be sequences of real numbers in $[0, 1]$. The higher-order Ishikawa iteration, will be the process, where $\{x_n\}_{n=0}^{\infty} \subset E$ is a sequence, defined by $x_{n+1} = (1 - \alpha_n)x_n + \alpha_n T^r u_n$, and $u_n = (1 - \beta_n)x_n + \beta_n T^r x_n$

Definition B.7 1

Let $\{\alpha_n\}_{n=0}^{\infty}$, $\{\beta_n\}_{n=0}^{\infty}$ and $\{\gamma_n\}_{n=0}^{\infty}$ be sequences of real numbers in $[0, 1]$. The higher-order Noor iteration, will be the process, where $\{x_n\}_{n=0}^{\infty} \subset E$ is a sequence, defined by $x_{n+1} = (1 - \alpha_n)x_n + \alpha_n T^r q_n$, and $q_n = (1 - \beta_n)x_n + \beta_n T^r v_n$, and $v_n = (1 - \gamma_n)x_n + \gamma_n T^r x_n$

Remark B.8 1

If $r = 1$ in the previous two definitions, we get the original Ishikawa and Noor iterations, respectively

Let (X, d) be a metric space. Recall from [Ampadu, Clement (2015):Generalization of Higher-Order Contraction Mapping Theorem. Unpublished] that a map $T : X \mapsto X$ is called a higher-order Banach contraction mapping if it holds that $d(T^r x, T^r y) \leq M\lambda^r d(x, y)$ for all $x, y \in X$ and any $r \in \mathbb{N}$, where $\lambda \in [0, 1)$ and $M \geq 1$ is the bound given by Proposition 4.1 contained in [Ezearn Fixed Point Theory and Applications (2015) 2015:88]. If $x := p$ is an r-fixed point of T, then there exists a modification on the bound given by Proposition 4.1[Ezearn Fixed Point Theory and Applications (2015) 2015:88] call it $M^\star$ such that $d(p, T^r y) \leq M^\star \lambda^r d(p, y)$. Throughout, we consider the contractive definition, $\|p - T^r y\| \leq M^\star \lambda^r \|p - y\|$, where $(E, \|\cdot\|)$ is a Banach space.

In the sequel we will need the following contained in [Berinde, V.: Iterative Approximation of Fixed Points. Editura Efemeride, Baia Mare, 2002]

Lemma B.9 1

Let $0 \leq \delta < 1$ be a real number, and $\{\epsilon_n\}$ a positive sequence satisfying $\epsilon_n \to 0$ as $n \to \infty$. Then, for any positive sequence $\{u_n\}$ satisfying $u_{n+1} \leq \delta u_n + \epsilon_n$, it follows that $u_n \to 0$ as $n \to \infty$

2.3 Main Results

Theorem B.1 1

Let $(E, \|\cdot\|)$ be a Banach space, and T be a self-map of E with r-fixed point p satisfying $\|p - T^r y\| \leq M^\star \lambda^r \|p - y\|$, where $M^\star$ is a certain modification on the bound given by Proposition 4.1[Ezearn Fixed Point Theory and Applications (2015) 2015:88] and $\lambda \in [0, 1)$. Let $\{x_n\}_{n=0}^\infty \subset E$ be a sequence satisfying Definition B.7 and converging to p, where $\{\alpha_n\}_{n=0}^\infty$, $\{\beta_n\}_{n=0}^\infty$ and $\{\gamma_n\}_{n=0}^\infty$ are sequences of real numbers in $[0, 1]$ such that $0 < \alpha_n \leq \alpha$, $0 < \beta_n \leq \beta$, and $0 < \gamma_n \leq \gamma$ for all n. Then the higher-order Noor iteration process is r-T-stable or r-stable with respect to T.

Proof of Theorem B.1 1

Suppose that $\{x_n\}_{n=0}^{\infty}$ converges to p. Let $\{y_n\}_{n=0}^{\infty}$ be an arbitrary sequence in E and put $\epsilon_n = \|y_{n+1} - (1-\alpha_n)y_n - \alpha_n T^r q_n\|$, for all $n = 0, 1, 2, \cdots$ and any $r \in \mathbb{N}$, where, $q_n = (1-\beta_n)y_n + \beta_n T^r v_n$, and $v_n = (1-\gamma_n)y_n + \gamma_n T^r y_n$. Assume that $\epsilon_n \to 0$ as $n \to \infty$, we now prove that $y_n \to p$ as $n \to \infty$. Observe from the triangle inequality, and the fact that $\|p - T^r y\| \leq M^\star \lambda^r \|p - y\|$, where $M^\star$ is a certain modification on the bound given by Proposition 4.1[Ezearn Fixed Point Theory and Applications (2015) 2015:88] and $\lambda \in [0,1)$, we have the following

$$\|y_{n+1} - p\| \leq \|y_{n+1} - (1-\alpha_n)y_n - \alpha_n T^r q_n\| + \|(1-\alpha_n)y_n + \alpha_n T^r q_n - p\|$$
$$= \epsilon_n + \|(1-\alpha_n)(y_n - p) + \alpha_n(T^r q_n - p)\|$$
$$\leq \epsilon_n + (1-\alpha_n)\|y_n - p\| + \alpha_n \|p - T^r q_n\|$$
$$\leq \epsilon_n + (1-\alpha_n)\|y_n - p\| + \alpha_n M^\star \lambda^r \|q_n - p\|$$

On the other hand

$$\|q_n - p\| = \|(1-\beta_n)y_n + \beta_n T^r v_n - [(1-\beta_n) + \beta_n]p\|$$
$$= \|(1-\beta_n)(y_n - p) + \beta_n(T^r v_n - p)\|$$
$$\leq (1-\beta_n)\|y_n - p\| + \beta_n \|p - T^r v_n\|$$
$$\leq (1-\beta_n)\|y_n - p\| + \beta_n M^\star \lambda^r \|v_n - p\|$$

From the previous two chain of inequalities, we deduce that,

$$\|y_n - p\| \leq \epsilon_n + [1 - (1 - M^\star \lambda^r)\alpha_n - \alpha_n \beta_n M^\star \lambda^r]\|y_n - p\|$$
$$+ \alpha_n \beta_n (M^\star \lambda^r)^2 \|v_n - p\|$$

On the other hand, we have,

$$\|v_n - p\| = \|(1-\gamma_n)y_n + \gamma_n T^r y_n - [(1-\gamma_n) + \gamma_n]p\|$$
$$= \|(1-\gamma_n)(y_n - p) + \gamma_n(T^r y_n - p)\|$$
$$\leq (1-\gamma_n)\|y_n - p\| + \gamma_n \|p - T^r y_n\|$$
$$\leq (1-\gamma_n)\|y_n - p\| + \gamma_n M^\star \lambda^r \|p - y_n\|$$
$$= (1 - \gamma_n + \gamma_n M^\star \lambda^r)\|y_n - p\|$$

Since $0 < \alpha_n \leq \alpha$, $0 < \beta_n \leq \beta$, and $0 < \gamma_n \leq \gamma$ for all n, then from the previous two inequalities immediately above, we deduce the following

$$\|y_{n+1} - p\| \leq \epsilon_n + [1 - (1 - M^\star \lambda^r)\alpha_n - \alpha_n \beta_n M^\star \lambda^r]\|y_n - p\|$$
$$+ \alpha_n \beta_n (M^\star \lambda^r)^2 [1 - \gamma_n + \gamma_n M^\star \lambda^r]\|y_n - p\|$$
$$\leq [1 - (1 - M^\star \lambda^r)\alpha - (1 - M^\star \lambda^r)\alpha\beta M^\star \lambda^r - (1 - M^\star \lambda^r)\alpha\beta\gamma(M^\star \lambda^r)^2]\|y_n - p\|$$
$$+ \epsilon_n$$

If we put $u_n := \|y_n - p\|$, and since $0 \leq \delta := 1 - (1 - M^\star \lambda^r)\alpha - (1 - M^\star \lambda^r)\alpha\beta M^\star \lambda^r - (1 - M^\star \lambda^r)\alpha\beta\gamma(M^\star \lambda^r)^2 < 1$, then from Lemma B.9, we deduce that $y_n \to p$ as $n \to \infty$, and the result follows.

Theorem B.2 1

Let $(E, \|\cdot\|)$ be a Banach space, and T be a self-map of E with r-fixed point p satisfying $\|p - T^r y\| \leq M^\star \lambda^r \|p - y\|$, where $M^\star$ is a certain modification on the bound given by Proposition 4.1[Ezearn Fixed Point Theory and Applications (2015) 2015:88] and $\lambda \in [0, 1)$. Let $\{x_n\}_{n=0}^\infty \subset E$ be a sequence satisfying Definition B.6 and converging to p, where $\{\alpha_n\}_{n=0}^\infty$ and $\{\beta_n\}_{n=0}^\infty$ are sequences of real numbers in $[0, 1]$ such that $0 < \alpha_n \leq \alpha$, $0 < \beta_n \leq \beta$ for all n. Then the higher-order Ishikawa iteration process is r-T-stable or r-stable with respect to T.

Proof of Theorem B.2 1

Suppose that $\{x_n\}_{n=0}^\infty$ converges to p. Let $\{y_n\}_{n=0}^\infty$ be an arbitrary sequence in E and put $\epsilon_n = \|y_{n+1} - (1 - \alpha_n)y_n - \alpha_n T^r u_n\|$ for all $n = 0, 1, 2, \cdots$ and any $r \in \mathbb{N}$, where $u_n = (1 - \beta_n)y_n + \beta_n T^r y_n$. Assume $\epsilon_n \to 0$ as $n \to \infty$, we show that $y_n \to p$ as $n \to \infty$. Observe from the triangle inequality, and the fact that $\|p - T^r y\| \leq M^\star \lambda^r \|p - y\|$, where $M^\star$ is a certain modification on the bound given by Proposition 4.1[Ezearn Fixed Point Theory and Applications (2015) 2015:88] and $\lambda \in [0, 1)$, we have the following,

$$\|y_{n+1} - p\| \leq \|y_{n+1} - (1 - \alpha_n)y_n - \alpha_n T^r u_n\| + \|(1 - \alpha_n)y_n + \alpha_n T^r u_n - p\|$$
$$= \epsilon_n + \|(1 - \alpha_n)y_n + \alpha_n T^r u_n - [(1 - \alpha_n) + \alpha_n]p\|$$
$$= \epsilon_n + \|(1 - \alpha_n)(y_n - p) + \alpha_n(T^r u_n - p)\|$$
$$\leq \epsilon_n + (1 - \alpha_n)\|y_n - p\| + \alpha_n \|p - T^r u_n\|$$
$$\leq \epsilon_n + (1 - \alpha_n)\|y_n - p\| + \alpha_n M^\star \lambda^r \|u_n - p\|$$

On the other hand

$$\|u_n - p\| = \|(1 - \beta_n)y_n + \beta_n T^r y_n - p\|$$
$$= \|(1 - \beta_n)y_n + \beta_n T^r y_n - [(1 - \beta_n) + \beta_n]p\|$$
$$= \|(1 - \beta_n)(y_n - p) + \beta_n(T^r y_n - p)\|$$
$$\leq (1 - \beta_n)\|y_n - p\| + \beta_n \|p - T^r y_n\|$$
$$\leq (1 - \beta_n)\|y_n - p\| + \beta_n M^\star \lambda^r \|p - y_n\|$$
$$= (1 - \beta_n + \beta_n M^\star \lambda^r)\|y_n - p\|$$

Since $0 < \alpha_n \leq \alpha$, $0 < \beta_n \leq \beta$ for all n, from the previous two chain of inequalities, we deduce the following

$$\|y_{n+1} - p\| \leq \epsilon_n + (1 - \alpha_n)\|y_n - p\|$$
$$+ \alpha_n M^\star \lambda^r (1 - \beta_n + \beta_n M^\star \lambda^r)\|y_n - p\|$$
$$= \epsilon_n + [(1 - \alpha_n) + \alpha_n M^\star \lambda^r (1 - \beta_n + \beta_n M^\star \lambda^r)]\|y_n - p\|$$
$$\leq \epsilon_n + [(1 - \alpha) + \alpha M^\star \lambda^r (1 - \beta + \beta M^\star \lambda^r)]\|y_n - p\|$$

If we put $u_n := \|y_n - p\|$, and since $0 \leq \delta := (1 - \alpha) + \alpha M^\star \lambda^r (1 - \beta + \beta M^\star \lambda^r) < 1$, then applying Lemma B.9, we conclude that $y_n \to p$ as $n \to \infty$, and the result follows

2.4 Exercises

> **Exercise B.1 1**
>
> Let (X, d) be a metric space, and let $Z \geq 1$ be the bound given by Proposition A.3 [Ampadu, Clement (2016). Characterization Theorems Inspired by the Hardy-Rogers Map I: Some Results in Metric Spaces, lulu.com. ISBN: 1365101185, 9781365101182]. A map $T : X \mapsto X$ is called a higher-order Hardy-Rogers type contraction if it satisfies $d(T^r x, T^r y) \leq Z\beta^r[d(x, Tx) + d(y, Ty) + d(y, Tx) + d(x, Ty) + d(x, y)]$ for all $x, y \in X$, where $\beta \in [0, \frac{1}{5})$. If $x := p$ is an r-fixed point of T, then the higher-order Hardy-Rogers type contraction can be written as $d(p, T^r y) \leq Z^\star \beta^r[d(y, Ty) + d(p, Ty) + 2d(y, p)]$, where $Z^\star$ is a certain modification on the bound given by Proposition A.3 [Ampadu, Clement (2016). Characterization Theorems Inspired by the Hardy-Rogers Map I: Some Results in Metric Spaces, lulu.com. ISBN: 1365101185, 9781365101182]. Using this contractive definition in Banach spaces, prove the following
>
> (a) Let $(E, \|\cdot\|)$ be a Banach space, and T be a self-map of E with r-fixed point p satisfying $\|p - T^r y\| \leq Z^\star \beta^r [\|y - Ty\| + \|p - Ty\| + 2\|y - p\|]$, where where $Z^\star$ is a certain modification on the bound given by Proposition A.3 [Ampadu, Clement (2016). Characterization Theorems Inspired by the Hardy-Rogers Map I: Some Results in Metric Spaces, lulu.com. ISBN: 1365101185, 9781365101182] and $\beta \in [0, \frac{1}{5})$. Let $\{x_n\}_{n=0}^\infty \subset E$ be a sequence satisfying Definition B.7 and converging to p, where $\{\alpha_n\}_{n=0}^\infty$, $\{\beta_n\}_{n=0}^\infty$ and $\{\gamma_n\}_{n=0}^\infty$ are sequences of real numbers in $[0, 1]$ such that $0 < \alpha_n \leq \alpha$, $0 < \beta_n \leq \beta$, and $0 < \gamma_n \leq \gamma$ for all n. Then the higher-order Noor iteration process is r-T-stable or r-stable with respect to T.
>
> (b) Let $(E, \|\cdot\|)$ be a Banach space, and T be a self-map of E with r-fixed point p satisfying $\|p - T^r y\| \leq Z^\star \beta^r [\|y - Ty\| + \|p - Ty\| + 2\|y - p\|]$, where $Z^\star$ is a certain modification on the bound given by Proposition A.3 [Ampadu, Clement (2016). Characterization Theorems Inspired by the Hardy-Rogers Map I: Some Results in Metric Spaces, lulu.com. ISBN: 1365101185, 9781365101182] and $\beta \in [0, \frac{1}{5})$. Let $\{x_n\}_{n=0}^\infty \subset E$ be a sequence satisfying Definition B.6 and converging to p, where $\{\alpha_n\}_{n=0}^\infty$ and $\{\beta_n\}_{n=0}^\infty$ are sequences of real numbers in $[0, 1]$ such that $0 < \alpha_n \leq \alpha$ and $0 < \beta_n \leq \beta$ for all n. Then the higher-order Ishikawa iteration process is r-T-stable or r-stable with respect to T.

> **Exercise B.2 1**
>
> Let (X,d) be a metric space. Recall from Exercise A.3 [Ampadu, Clement (2016). Characterization Theorems Inspired by the Hardy-Rogers Map I: Some Results in Metric Spaces, lulu.com. ISBN: 1365101185, 9781365101182], there exist $J \geq 1$ and $\Gamma \in [0, \frac{1}{3})$ such that $d(T^r x, T^r y) \leq J\Gamma^r[d(x,Tx) + d(y,Ty) + d(x,y)]$, and we say $T : X \mapsto X$ is a higher-order Reich type contraction mapping. If $x := p$ is an r-fixed point of T, then the higher-order Reich type contraction mapping in Banach spaces can be written as
>
> $$\|p - T^r y\| \leq J^\star \Gamma^r [\|y - Ty\| + \|p - y\|] \qquad (2.1)$$
>
> where $J^\star$ is a certain modification on J. Using this contractive definition prove the following
>
> (a) Let $(E, \|\cdot\|)$ be a Banach space, and T be a self-map of E with r-fixed point p satisfying (2.1). Let $\{x_n\}_{n=0}^\infty \subset E$ be a sequence satisfying Definition B.7 and converging to p, where $\{\alpha_n\}_{n=0}^\infty$, $\{\beta_n\}_{n=0}^\infty$ and $\{\gamma_n\}_{n=0}^\infty$ are sequences of real numbers in $[0,1]$ such that $0 < \alpha_n \leq \alpha$, $0 < \beta_n \leq \beta$, and $0 < \gamma_n \leq \gamma$ for all n. Then the higher-order Noor iteration process is r-T-stable or r-stable with respect to T.
>
> (b) Let $(E, \|\cdot\|)$ be a Banach space, and T be a self-map of E with r-fixed point p satisfying (2.1). Let $\{x_n\}_{n=0}^\infty \subset E$ be a sequence satisfying Definition B.6 and converging to p, where $\{\alpha_n\}_{n=0}^\infty$ and $\{\beta_n\}_{n=0}^\infty$ are sequences of real numbers in $[0,1]$ such that $0 < \alpha_n \leq \alpha$ and $0 < \beta_n \leq \beta$ for all n. Then the higher-order Ishikawa iteration process is r-T-stable or r-stable with respect to T.

2.5 References

(1) Noor, M. A.: New approximations schemes for general variational inequalities. J. Math. Anal. Appl. 251 (2000), 217–299

(2) S. ISHIKAWA, Fixed points by a new iteration method, Proc. Amer. Math. Soc. 44 (1974),147-150

(3) Ampadu, Clement (2015):Generalization of Higher-Order Contraction Mapping Theorem. Unpublished

(4) Ezearn Fixed Point Theory and Applications (2015) 2015:88

(5) Berinde, V.: Iterative Approximation of Fixed Points. Editura Efemeride, Baia Mare, 2002

(6) Ampadu, Clement (2016). Characterization Theorems Inspired by the Hardy-Rogers Map I: Some Results in Metric Spaces, lulu.com. ISBN: 1365101185, 9781365101182

Chapter 3

Stability of the Higher-Order Hardy-Rogers Type Contraction in Multiplicative Cone Metric Space

3.1 Brief Summary

> **Abstract C.1 1**
>
> Recall from Theorem A.1 [Ampadu, Clement (2016). Characterization Theorems Inspired by the Hardy-Rogers Map II: Some Results in Cone Metric Spaces, lulu.com. ISBN: 1365109917, 9781365109911] that we gave the higher-order Hardy-Rogers type mapping theorem in cone metric space. In the present chapter theorems on convergence and stability in complete multiplicative cone metric space are presented

3.2 Preliminaries

> **Notation C.1 1**
>
> E will denote a real Banach space

> **Remark C.2 1**
>
> We assume that $P \subset E$ with $int(P) \neq \emptyset$, where $int(P)$ denotes the interior of P

> **Definition C.3 1**
>
> We will say P is a multiplicative cone if the following hold
>
> (a) P is closed, nonempty, and $P \neq \{1\}$
>
> (b) If $a, b \in \mathbb{R}$, $a, b \geq 1$, then $x^a \cdot y^b \in P$
>
> (c) If $x \in P$ and $x \in \frac{1}{P}$, then $x = 1$

> **Notation C.4 1**
>
> $\leq$ will denote a partial ordering with respect to P and will be defined as $x \leq y$ iff $\frac{y}{x} \in P$.
> We shall write $x < y$ iff $x \leq y$ but $x \neq y$, and $x \ll y$ iff $\frac{y}{x} \in int(P)$

Remark C.5 1

If $a \leq b$ and $c \leq d$, then, $ac \leq ad$, and for $\lambda \in \mathbb{R}$, $\lambda \geq 0$, $a^\lambda \leq b^\lambda$

Definition C.6 1

The cone P will be called multiplicative normal if there is a number $K > 0$ such that for all $x, y \in E$, $1 \leq x \leq y$ implies $\|x\| \leq \|y\|^K$. The least positive number satisfying $\|x\| \leq \|y\|^K$ will be called the multiplicative normal constant of P

Remark C.7 1

We always assume that E is a real Banach space and $P \subset E$ with $int(P) \neq \emptyset$, and $\leq$ is a partial ordering with respect to P

Definition C.8 1

Let X be a nonempty set. Suppose the mapping $m : X \times X \mapsto E$ satisfies

(a) $1 \leq m(x, y)$ for all $x, y \in X$ and $m(x, y) = 1$ iff $x = y$

(b) $m(x, y) = m(y, x)$ for all $x, y \in X$

(c) $m(x, y) \leq m(x, z) \cdot m(z, y)$ for all $x, y, z \in X$

Then we say m is a multiplicative cone metric on X and (X, m) will be called a multiplicative cone metric space

Definition C.9 1

Let (X, m) be a multiplicative cone metric space. Let $\{x_n\}$ be a sequence in X and $x \in X$. If for every $c \in E$ with $1 \ll c$ there is N such that for all $n > N$, $m(x_n, x) \ll c$, then $\{x_n\}$ is said to be multiplicative convergent

Definition C.10 1

Let (X, m) be a multiplicative cone metric space, and let $\{x_n\}$ be a sequence in X. If for any $c \in E$ with $1 \ll c$, there is N such that for all $n, k > N$, $m(x_n, x_k) \ll c$, then $\{x_n\}$ is said to be a multiplicative Cauchy sequence in X.

Definition C.11 1

If every multiplicative Cauchy sequence is multiplicative convergent in (X, m), then (X, m) is said to be a complete multiplicative cone metric space

Definition C.12 1

Let (X, m) be a multiplicative cone metric space, and T a selfmap of X. Let $x_{n+1} = f(T^r, x_n)$ for all $n = 0, 1, 2, \cdots$ and any $r \in \mathbb{N}$, where f is some function, be some iteration procedure. Suppose for any $r \in \mathbb{N}$, that, $F(T^r)$, the set of r-fixed points of T is nonempty, and that x_n converges to a point $p \in F(T^r)$. Let $\{y_n\} \subset X$, and define $\epsilon_n = m(y_{n+1}, f(T^r, y_n))$. If $\epsilon_n \to 1$ as $n \to \infty$ implies $y_n \to p$ as $n \to \infty$, then we will say $x_{n+1} = f(T^r, x_n)$ is r-T-stable or r-stable with respect to T.

3.3 Main Results

Theorem C.1 1

Let (X, m) be a complete multiplicative cone metric space, and T be a self-map of X such that $m(T^r x, T^r y) \leq [m(x, Tx) \cdot m(y, Ty) \cdot m(x, Ty) \cdot m(y, Tx) \cdot m(x, y)]^{Z\beta^r}$ for all $x, y \in X$, where $\beta \in [0, \frac{1}{5})$ and $Z \geq 1$ is the bound given by Proposition A.3 [Ampadu, Clement (2016): Characterization Theorems Inspired by the Hardy-Rogers Map I: Some Results in Metric Spaces, lulu.com. ISBN: 1365101185, 9781365101182], then T has a unique r-fixed point in X. Moreover, if $\{x_n\}$ is generated by a so-called higher-order Picard iteration process: $x_0 \in X$, $x_n = T^r x_{n-1}$ for all $n = 0, 1, 2, \cdots$ and any $r \in \mathbb{N}$, then $\{x_n\}$ converges to an r-fixed point of T

Proof of Theorem C.1 1

Take $x = x_{n-1}$ and $y = x_n$ in the contractive definition of the theorem, then we have, $m(T^r x_{n-1}, T^r x_n) \leq [m(x_{n-1}, Tx_{n-1}) \cdot m(x_n, Tx_n) \cdot m(x_{n-1}, Tx_n) \cdot m(x_n, Tx_{n-1}) \cdot m(x_{n-1}, x_n)]^{Z\beta^r}$. From which one deduces that $m(x_n, x_{n+1}) \leq [m(x_{n-1}, x_n) \cdot m(x_n, x_{n+1}) \cdot m(x_{n-1}, x_{n+1}) \cdot m(x_n, x_n) \cdot m(x_{n-1}, x_n)]^{Z\beta^r}$. Since $m(x_n, x_n) = 1$, an application of the multiplicative triangle inequality implies that $m(x_n, x_{n+1}) \leq [m(x_{n-1}, x_n)^3 \cdot m(x_n, x_{n+1})^2]^{Z\beta^r}$, from which one deduces that $m(x_n, x_{n+1}) \leq m(x_{n-1}, x_n)^h$, where $h = \frac{3Z\beta^r}{1-2Z\beta^r}$. By induction, we have, $m(x_{n+1}, x_n) \leq m(x_1, x_0)^{h^n}$. Now for $n > k$, we have, $m(x_n, x_k) \leq m(x_n, x_{n-1}) \cdot m(x_{n-1}, x_{n-2}) \cdot \cdots \cdot m(x_{k+1}, x_k) \leq m(x_1, x_0)^{h^{n-1}+h^{n-2}+\cdots+h^k} \leq m(x_1, x_0)^{\frac{h^k}{1-h}}$. For a given $c \in E$ with $1 \ll c$, that is, $c \in int(P)$, there exist a multiplicative ball, $B(0, r)$ such that $c \cdot B(0, r) \subseteq P$, where $B(0, r) = \{x \in E, \|x\| \leq r, r > 1\}$, but there exists a positive number N such that $m(x_1, x_0)^{\frac{h^k}{1-h}} \in B(0, r)$ for all $k > N$, therefore we have, $m(x_1, x_0)^{\frac{h^k}{1-h}} \ll c$, for all $k > N$. It follows that, $m(x_n, x_k) \leq m(x_1, x_0)^{\frac{h^k}{1-h}} \ll c$, thus $\{x_n\}$ is a multiplicative Cauchy sequence and since X is multiplicative complete, there exists $p \in X$ such that $x_n \to p$ as $n \to \infty$. Now observe that, $m(p, T^r p) \leq m(p, x_{n+1}) \cdot m(x_{n+1}, T^r p) = m(p, x_{n+1}) \cdot m(T^r x_n, T^r p) \leq m(p, x_{n+1}) \cdot m(x_n, p)^h \to 1$ as $n \to \infty$. It follows that $T^r p = p$

Lemma C.2 1

Let (X, m) be a multiplicative cone metric space with respect to a multiplicative normal cone P of E. Let $\{a_n\}$ and $\{\epsilon_n\}$ be sequences in E satisfying $a_n \leq a_{n-1}^h \cdot \epsilon_n$, where $h \in [0, 1)$, and $\epsilon_n \to 1$ as $n \to \infty$, then $a_n \to 1$ as $n \to \infty$

Proof of Lemma C.2 1

Notice that

$$\begin{aligned}
a_n &\leq a_{n-1}^h \cdot \epsilon_n \\
&\leq a_{n-2}^{h^2} \cdot \epsilon_{n-1}^h \cdot \epsilon_n \\
&\leq \\
&\vdots \\
&\leq a_0^{h^n} \cdot \epsilon_1^{h^{n-1}} \cdot \epsilon_2^{h^{n-2}} \cdot \ldots \cdot \epsilon_{n-1}^h \cdot \epsilon_n
\end{aligned}$$

It follows that

$$\begin{aligned}
\|a_n\| &\leq \|a_0^{h^n} \cdot \epsilon_1^{h^{n-1}} \cdot \epsilon_2^{h^{n-2}} \cdot \ldots \cdot \epsilon_{n-1}^h \cdot \epsilon_n\|^K \\
&\leq (\|a_0\|^{h^n} \cdot \|\epsilon_1\|^{h^{n-1}} \cdot \|\epsilon_2\|^{h^{n-2}} \cdot \ldots \cdot \|\epsilon_{n-1}\|^h \cdot \|\epsilon_n\|)^K \\
&\leq \|a_0\|^{h^n K} \cdot \sup\{\|\epsilon_n\|\}^{K(h^{n-1}+h^{n-2}+\cdots+h+1)} \to 1
\end{aligned}$$

It follows that $a_n \to 1$ as $n \to \infty$

Theorem C.3 1

Let (X, m) be a complete multiplicative cone metric space with respect to a multiplicative normal cone P. Let T be a self-map on X such that $m(T^r x, T^r y) \leq [m(x, Tx) \cdot m(y, Ty) \cdot m(x, Ty) \cdot m(y, Tx) \cdot m(x, y)]^{Z\beta^r}$ for all $x, y \in X$, where $\beta \in [0, \frac{1}{5})$ and $Z \geq 1$ is the bound given by Proposition A.3 [Ampadu, Clement (2016): Characterization Theorems Inspired by the Hardy-Rogers Map I: Some Results in Metric Spaces, lulu.com. ISBN: 1365101185, 9781365101182]. Assume that T enjoys an r-fixed point. Let $\{x_n\}$ be a sequence generated by the so-called higher-order Picard iteration process: $x_0 \in X$, $x_n = T^r x_{n-1}$ for all $n = 0, 1, 2, \cdots$ and any $r \in \mathbb{N}$, then the higher-order Picard iteration process is r-stable with respect to T or r-T-stable.

Proof of Theorem C.3 1

Let $\{y_n\}$ be an arbitrary sequence and set $\epsilon_n = m(y_{n+1}, T^r y_n)$. Assume $\epsilon_n \to 1$ as $n \to \infty$, we must show $y_n \to p$ as $n \to \infty$, where p is an r-fixed point of T. Notice that $m(y_{n+1}, p) \leq m(y_{n+1}, T^r y_n) \cdot m(T^r y_n, p) = m(y_{n+1}, T^r y_n) \cdot m(T^r y_n, T^r p) \leq \epsilon_n \cdot m(y_n, p)^h$, where $0 \leq h := \frac{3Z\beta^r}{1-2Z\beta^r} < 1$. Therefore, applying Lemma C.2, we deduce that $m(y_n, p) \to 1$ as $n \to \infty$, thus, $y_n \to p$ as $n \to \infty$

3.4 Exercises

Exercise C.1 1

Prove the following: Let (X, m) be a complete multiplicative cone metric space, and T be a self-map of X such that $m(T^r x, T^r y) \leq [m(x, Tx) \cdot m(y, Ty) \cdot m(x, y)]^{J\Gamma^r}$ for all $x, y \in X$, where $\Gamma \in [0, \frac{1}{3})$ and $J \geq 1$ is the bound given by Exercise A.3 [Ampadu, Clement (2016): Characterization Theorems Inspired by the Hardy-Rogers Map I: Some Results in Metric Spaces, lulu.com. ISBN: 1365101185, 9781365101182], then T has a unique r-fixed point in X. Moreover, if $\{x_n\}$ is generated by a so-called higher-order Picard iteration process: $x_0 \in X$, $x_n = T^r x_{n-1}$ for all $n = 0, 1, 2, \cdots$ and any $r \in \mathbb{N}$, then $\{x_n\}$ converges to an r-fixed point of T

> **Exercise C.2 1**
>
> Prove the following: Let (X, m) be a complete multiplicative cone metric space with respect to a multiplicative normal cone P. Let T be a self-map on X such that $m(T^r x, T^r y) \leq [m(x, Tx) \cdot m(y, Tx) \cdot m(x, y)]^{J\Gamma^r}$ for all $x, y \in X$, where $\Gamma \in [0, \frac{1}{3})$ and $J \geq 1$ is the bound given by Exercise A.3 [Ampadu, Clement (2016): Characterization Theorems Inspired by the Hardy-Rogers Map I: Some Results in Metric Spaces, lulu.com. ISBN: 1365101185, 9781365101182]. Assume that T enjoys an r-fixed point. Let $\{x_n\}$ be a sequence generated by the so-called higher-order Picard iteration process: $x_0 \in X$, $x_n = T^r x_{n-1}$ for all $n = 0, 1, 2, \cdots$ and any $r \in \mathbb{N}$, then the higher-order Picard iteration process is r-stable with respect to T or r-T-stable.

3.5 References

(1) Ampadu, Clement (2016). Characterization Theorems Inspired by the Hardy-Rogers Map II: Some Results in Cone Metric Spaces, lulu.com. ISBN: 1365109917, 9781365109911

(2) Ampadu, Clement (2016): Characterization Theorems Inspired by the Hardy-Rogers Map I: Some Results in Metric Spaces, lulu.com. ISBN: 1365101185, 9781365101182

Chapter 4

Stability of the Higher-Order Jungck-Kannan Mapping Operator in b-Metric Space

4.1 Brief Summary

> **Abstract D.1 1**
>
> Recall that the Kannan mapping theorem was given in [R. Kannan. Some results on fixed points II. American Mathematical Monthly, 76 (1969) 405-408]. On the other hand the Jungck-Mann iteration scheme was given in [S. L. Singh, C. Bhatnagar and S. N. Mishra, "Stability of Jungck-type iterative procedures," International Journal of Mathematics and Mathematical Sciences, vol. 2005, pp. 3035-3043, 2005]. In this chapter, we introduce a generalization of the Jungck-Mann iteration scheme, which we call the higher-order Jungck-Mann iteration scheme and use it to obtain a stability result for a generalization of the Kannan mapping theorem corresponding to a pair of mappings with a common fixed point in b-metric space, hence the title of this chapter

4.2 Preliminaries

> **Definition D.1 1**
>
> Let (X, d) be a complete metric space, and $T : X \mapsto X$. The point which satisfies $T^r x = x$ for any $r \in \mathbb{N}$ will be called the r-fixed point of T

> **Remark D.2 1**
>
> A 1-fixed point of T is a fixed point of T

> **Definition D.3 1**
>
> Let $\{x_n\}_{n=0}^{\infty} \subset X$ be a sequence generated by the iteration procedure involving the operator T, then the iterative procedure defined by $x_{n+1} = f(T^r, x_n) = T^r x_n$, $n = 0, 1, \cdots$, and any $r \in \mathbb{N}$ will be called the higher-order Picard iterative procedure

CHAPTER 4. STABILITY OF THE HIGHER-ORDER JUNGCK-KANNAN MAPPING OPERATOR IN B-METRIC SPACE

Definition D.4 1

If $f(T^r, x_n) = (1 - \alpha_n)x_n + \alpha_n T^r x_n$, then for $x_0 \in X$ and $\{x_n\}_{n=0}^{\infty} \subset [0,1]$, the iterative procedure defined by $x_{n+1} = (1 - \alpha_n)x_n + \alpha_n T^r x_n$, $n = 0, 1, \cdots$, and any $r \in \mathbb{N}$ will be called the higher-order Mann iterative procedure

Remark D.5 1

If $r = 1$ in the previous definition, then the higher-order Mann iterative procedure gives the usual Mann iterative procedure [W. R. Mann, "Mean value methods in iteration," Proceedings of the American Mathematical Society, vol. 4, pp. 506-510, 1953]

Remark D.6 1

If $\{\alpha_n\} = 1$ in the previous definition, then the higher-order Mann iterative procedure gives the higher-order Picard iterative procedure. On the other hand, if $r = 1$ and $\{\alpha_n\} = 1$ in the previous definition, then the higher-order Mann iterative procedure gives the usual Picard iterative procedure

Definition D.7 1

Let Y be an arbitrary non-empty set and (X, d) be a metric space. Let $S, T : Y \mapsto X$ and $T^r(Y) \subset S^r(Y)$ for any $r \in \mathbb{N}$ and for some $x_0 \in Y$. Consider $S^r x_{n+1} = f(T^r, x_n)$, $n = 0, 1, 2, \cdots$ and any $r \in \mathbb{N}$. For $Y = X$ and $f(T^r, x_n) = T^r x_n$, the iterative procedure defined by $S^r x_{n+1} = T^r x_n$, $n = 0, 1, 2, \cdots$ and any $r \in \mathbb{N}$ will be called the higher-order Jungck iteration scheme

Remark D.8 1

If $r = 1$ in the above definition, then the higher-order Jungck iteration scheme gives the usual Jungck iteration scheme in the literature. On putting $Y = X$ and $S = id$, the identity mapping on X in the above definition, we get the the higher-order Picard iterative procedure. If in addition, we assume $r = 1$, then we get the usual Picard iterative procedure

Definition D.9 1

Let Y be an arbitrary non-empty set and (X, d) be a metric space. Let $S, T : Y \mapsto X$ and $T^r(Y) \subset S^r(Y)$ for any $r \in \mathbb{N}$ and for some $x_0 \in Y$. Consider $S^r x_{n+1} = f(T^r, x_n)$, $n = 0, 1, 2, \cdots$ and any $r \in \mathbb{N}$. For $Y = X$ and $f(T^r, x_n) = (1 - \alpha_n)S^r x_n + \alpha_n T^r x_n$, the iterative procedure defined by $S^r x_{n+1} = (1 - \alpha_n)S^r x_n + \alpha_n T^r x_n$, $n = 0, 1, 2, \cdots$ and any $r \in \mathbb{N}$, where $\{\alpha_n\}_{n=0}^{\infty} \subset [0,1]$, will be called the higher-order Jungck-Mann iteration scheme

Remark D.10 1

If $r = 1$ in the above definition, then the higher-order Jungck-Mann iteration scheme gives the usual Jungck-Mann iteration scheme in the literature. On putting $Y = X$ and $S = id$, the identity mapping on X in the above definition, we get the the higher-order Mann iterative procedure. If in addition, we assume $r = 1$, then we get the usual Mann iterative procedure

Taking inspiration from Harder and Hicks [A. M. Harder and T. L. Hicks, A stable iteration procedure for nonexpansive mappings, Math. Japonica, vol. 33, pp. 687-692, 1988; A. M. Harder and T. L. Hicks, Stability results for fixed point iteration procedures, Math. Japonica, vol. 33, pp. 693-706, 1988] we introduce the following

CHAPTER 4. STABILITY OF THE HIGHER-ORDER JUNGCK-KANNAN MAPPING OPERATOR IN B-METRIC SPACE

> **Definition D.11 1**
>
> An iterative procedure $x_{n+1} = f(T^r, x_n)$ for all $n = 0, 1, 2, \cdots$ and any $r \in \mathbb{N}$ will be called r-T-stable or r-stable with respect to a mapping T if $\{x_n\}$ converges to a r-fixed point q of T and whenever $\{y_n\}$ is a sequence in X with $d(y_{n+1}, f(T^r, y_n)) \to 0$ as $n \to \infty$ we have $y_n \to q$ as $n \to \infty$.

Taking inspiration from [S. L. Singh, C. Bhatnagar and S. N. Mishra, "Stability of Jungck-type iterative procedures," International Journal of Mathematics and Mathematical Sciences, vol. 2005, pp. 3035-3043, 2005] we introduce the following

> **Definition D.12 1**
>
> Let $S, T : Y \mapsto X$, $T^r(Y) \subset S^r(Y)$ for any $r \in \mathbb{N}$, and z be a r-coincidence point of T and S, that is, $S^r z = T^r z = p$ (say) for any $r \in \mathbb{N}$. Fix $x_0 \in Y$, and let the sequence $\{S^r x_n\}$ converge to p for any $r \in \mathbb{N}$. Let $\{S^r y_n\} \subset X$ be an arbitrary sequence, and set $\epsilon_n = d(S^r y_{n+1}, f(T^r, y_n))$, $n = 0, 1, 2, \cdots$ and any $r \in \mathbb{N}$, then the iterative procedure $f(T^r, x_n)$ will be called $r - (S,T)$ stable or r-stable with respect to (S,T) iff $\epsilon_n \to 0$ as $n \to \infty$ implies $S^r y_n \to p$ as $n \to \infty$.

The concept of b-metric space was introduced in [S. Czerwik, "Nonlinear set-valued contraction mappings in b-metric spaces," Atti Del Seminario Matematico E Fisico Universita Di Modena, vol. 46, pp. 263-276, 1998], and we can state it as follows

> **Definition D.13 1**
>
> Let X be a set and $s \geq 1$ be a real number. A function $d : X \times X \mapsto \mathbb{R}^+$ is said to be a b-metric if it satisfies the following for all $x, y, z \in X$
>
> (a) $d(x, y) = 0$ iff $x = y$
>
> (b) $d(x, y) = d(y, x)$
>
> (c) $d(x, z) \leq s[d(x, y) + d(y, z)]$
>
> A pair (X, d) is called a b-metric space

Recall [R. Kannan. Some results on fixed points II. American Mathematical Monthly, 76 (1969) 405-408], that a map $S : X \mapsto X$ is called a Kannan mapping if $d(Sx, Sy) \leq \lambda[d(x, Sx) + d(y, Sy)]$ for all $x, y \in X$ and $\lambda \in [0, \frac{1}{2})$. If we replace the identity map with a continuous map, $T : X \mapsto X$ (say), then we get the following generalization of the Kannan mapping, namely, $d(Sx, Sy) \leq \lambda[d(Tx, Sx) + d(Ty, Sy)]$ for all $x, y \in X$ and $\lambda \in [0, \frac{1}{2})$, and we can obtain a common fixed point theorem in the sense of Jungck [G. Jungck, "Commuting mappings and fixed points," Am. Math. Mon., vol. 83, pp. 261–263, 1976]. Such generalizations of the Kannan mapping, we term Jungck-Kannan contractions.

> **Definition D.14 1**
>
> Let (X, d) be a metric space and let the self-maps $S, T : X \to X$ satisfy $d(S^r x, S^r y) \leq \sum_{q=0}^{r-1} c_q [d(T^{q+1} x, S^{q+1} x) + d(T^{q+1} y, S^{q+1} y)]$ for all x, y in X and $r \in \mathbb{N}$. If $0 \leq c_q < \frac{1}{2}$ for all $0 \leq q \leq r - 1$, then, $S : X \to X$ is said to be an rth order Jungck-Kannan contraction mapping.

> **Proposition D.15 1**
>
> Let (X,d) be a metric space, and let S be an rth-order Jungck-Kannan mapping on X. For every pair $x \neq y \in X$, define
>
> $$Z := Z(x,y) = \max_{0 \leq v \leq r-1} \beta^{-v} \frac{d(S^v x, S^v y)}{d(Tx, Sx) + d(Ty, Sy)}$$
>
> then
>
> $$Z = \max_{n \in 0 \cup \mathbb{N}} \beta^{-n} \frac{d(S^n x, S^n y)}{d(Tx, Sx) + d(Ty, Sy)}$$
>
> where $\beta \in [0, \frac{1}{2})$

Now we give the higher-order version of the Jungck-Kannan contraction as follows

> **Definition D.16 1**
>
> Let (X,d) be a metric space, and $S, T : X \mapsto X$ satisfy $d(S^r x, S^r y) \leq Z\beta^r[d(Tx, Sx) + d(Ty, Sy)]$ for all $x, y \in X$, where $Z \geq 1$ is given by the previous proposition and $\beta \in [0, \frac{1}{2})$, then we say S is a higher-order Jungck-Kannan contraction

The main goal in the next section is to prove r-stability for higher-order Jungck-Mann iteration procedure in the setting of b-metric space using the higher-order Jungck-Kannan contractive condition, that is, the previous definition

4.3 Main Results

> **Theorem D.1 1**
>
> Let (X,d) be a b-metric space and S,T be maps on an arbitrary set Y with values in X such that $T^r(Y) \subseteq S^r(Y)$ and $S^r(Y)$ or $T^r(Y)$ is a complete subspace of X for any $r \in \mathbb{N}$. Let z be a r-coincidence point of T and S, that is, $S^r z = T^r z = p$. Let $x_0 \in Y$, and let the sequence $\{S^r x_n\}$ for any $r \in \mathbb{N}$, generated by $S^r x_{n+1} = (1-\alpha_n) S^r x_n + \alpha_n T^r x_n$, $n = 0, 1, 2, \cdots$, and any $r \in \mathbb{N}$, where $\{\alpha_n\}$ for $n = 0, 1, 2, \cdots$, satisfy
>
> (i) $\alpha_0 = 1$
>
> (ii) $0 \leq \alpha_n < 1$, $n = 1, 2, \cdots$
>
> (iii) $\sum \alpha_n = \infty$
>
> (iv) $\sum_{j=0}^{n} \alpha_j \prod_{i=j+1}^{n} (1 - \alpha_i + Z\beta^r \alpha_i)$
>
> converge to p. Let $\{S^r y_n\} \subset X$ and define $\epsilon_n = d(S^r y_{n+1}, (1-\alpha_n) S^r y_n + \alpha_n T^r y_n)$, $n = 0, 1, 2, \cdots$. If the pair (S, T) satisfy Definition D.16 for all $x, y \in Y$. Then
>
> (I) $d(p, S^r y_{n+1}) \leq d(p, S^r x_{n+1}) + s^2 \prod_{i=0}^{n}(1-\alpha_i) d(S^r x_0, S^r y_0) + Z\beta^r \sum_{j=0}^{n} s^{n-i} \alpha_j \prod_{i=j+1}^{n}(1-\alpha_i)[d(Sx_i, Tx_i) + d(Sy_i, Ty_i)] + \sum_{j=0}^{n} s^{n-i} \alpha_j \prod_{i=j+1}^{n}(1-\alpha_j)\epsilon_j$
>
> (II) $S^r y_n \to p$ as $n \to \infty$ iff $\epsilon_n \to 0$ as $n \to \infty$

Proof of Theorem D.1 1

First observe that

$$\begin{aligned}
d(p, S^r y_{n+1}) &\leq s[d(p, S^r x_{n+1}) + d(S^r x_{n+1}, S^r y_{n+1})] \\
&\leq s[d(p, S^r x_{n+1}) + d((1-\alpha_n)S^r x_n + \alpha_n T^r x_n, S^r y_{n+1})] \\
&\leq sd(p, S^r x_{n+1}) + s[d((1-\alpha_n)S^r x_n + \alpha_n T^r x_n, (1-\alpha_n)S^r y_n + \alpha_n T^r y_n) \\
&\quad + d((1-\alpha_n)S^r y_n + \alpha_n T^r y_n, S^r y_{n+1})] \\
&\leq sd(p, S^r x_{n+1}) + s[(1-\alpha_n)d(S^r x_n, S^r y_n) + \alpha_n d(T^r x_n, T^r y_n)] + s\epsilon_n \\
&\leq sd(p, S^r x_{n+1}) + s(1-\alpha_n)d(S^r x_n, S^r y_n) \\
&\quad + sZ\beta^r \alpha_n[d(Sx_n, Tx_n) + d(Sy_n, Ty_n)] + s\epsilon_n
\end{aligned}$$

On the other hand notice that

$$\begin{aligned}
d(S^r x_n, S^r y_n) &\leq s[d((1-\alpha_{n-1})S^r x_{n-1} + \alpha_{n-1}T^r x_{n-1}, (1-\alpha_n)S^r y_{n-1} + \alpha_{n-1}T^r y_{n-1}) \\
&\quad + d((1-\alpha_{n-1})S^r y_{n-1} + \alpha_{n-1}T^r y_{n-1}, S^r y_n)] \\
&\leq s[(1-\alpha_{n-1})d(S^r x_{n-1}, S^r y_{n-1}) + \alpha_{n-1}d(T^r x_{n-1}, T^r y_{n-1})] + s\epsilon_{n-1} \\
&\leq s(1-\alpha_{n-1})d(S^r x_{n-1}, S^r y_{n-1}) \\
&\quad + sZ\beta^r \alpha_{n-1}[d(Sx_{n-1}, Tx_{n-1}) + d(Sy_{n-1}, Ty_{n-1})] + s\epsilon_{n-1}
\end{aligned}$$

From the two "chain of inequalities" immediately above, we deduce the following

$$\begin{aligned}
d(p, S^r y_{n+1}) &\leq sd(p, S^r x_{n+1}) + s^2(1-\alpha_n)(1-\alpha_{n-1})d(S^r x_{n-1}, S^r y_{n-1}) \\
&\quad + s^2(1-\alpha_n)Z\beta^r \alpha_{n-1}[d(Sx_{n-1}, Tx_{n-1}) + d(Sy_{n-1}, Ty_{n-1})] \\
&\quad + s^2(1-\alpha_n)\epsilon_{n-1} + sZ\beta^r \alpha_n[d(Sx_n, Tx_n) + d(Sy_n, Ty_n)] + s\epsilon_n
\end{aligned}$$

This process when repeated $(n-1)$ times, yields (I). To prove (II), suppose that $S^r y_n \to p$ as $n \to \infty$, then,

$$\begin{aligned}
\epsilon_n &= d(S^r y_{n+1}, (1-\alpha_n)S^r y_n + \alpha_n T^r y_n) \\
&\leq d(S^r y_{n+1}, p) + d(p, (1-\alpha_n)S^r y_n + \alpha_n T^r y_n) \\
&\leq d(S^r y_{n+1}, p) + d((1-\alpha_n + \alpha_n)p, (1-\alpha_n)S^r y_n + \alpha_n T^r y_n) \\
&\leq d(S^r y_{n+1}, p) + (1-\alpha_n)d(p, S^r y_n) + \alpha_n d(T^r p, T^r y_n) \\
&\leq d(S^r y_{n+1}, p) + (1-\alpha_n)d(p, S^r y_n) + \alpha_n Z\beta^r[d(Sp, Tp) + d(Sy_n, Ty_n)] \to 0
\end{aligned}$$

Now suppose that $\epsilon_n \to 0$ as $n \to \infty$. Let A denote the lower triangular matrix with entries $\alpha_{nj} = \alpha_j \prod_{i=j+1}^{n}(1-\alpha_i)$, then A is multiplicative, so that, $Z\beta^r \sum_{j=0}^{n} s^{n-i}\alpha_j \prod_{i=j+1}^{n}(1-\alpha_i)[d(Sx_i, Tx_i) + d(Sy_i, Ty_i)] \to 0$ as $n \to \infty$ and $\sum_{j=0}^{n} s^{n-1}\alpha_j \prod_{i=j+1}^{n}(1-\alpha_i)\epsilon_j \to 0$ as $n \to \infty$. Finally, condition (iii) of the iterative scheme implies $\prod_{i=j+1}^{n}(1-\alpha_i + Z\beta^r \alpha_i) \to 0$ as $n \to \infty$. Hence $S^r y_n \to p$ as $n \to \infty$

Remark D.2 1

If we take $s = 1$ in the above theorem, then we get the corresponding result in metric space

Remark D.3 1

If we put $Y = X$ and $S = id$, the identity map on X in the above theorem, then we get the corresponding result for the higher-order Mann iterative procedure

Finally we have the following

> **Example D.4 1**
>
> Let $f^r x = x - 2r$ for any $r \in \mathbb{N}$. Notice that we can write $f^r x = 2r(\frac{x}{2r} - 1)$. Put $S^r x = \frac{x}{2r}$ and $T^r x = 1$ and notice that $f^r x = 2r(S^r x - T^r x)$. Now observe that $S^r(2r) = T^r(2r) = 1$, and $2r$ is the zero of $f^r x = x - 2r$. Let $\{\alpha_n\}$ be a sequence converging to 1, then the higher-order Jungck-Mann iterative scheme implies, $\lim_{n \to \infty} S^r x_{n+1} = \lim_{n \to \infty} T^r x_n = 1$, and by the continuity of the functions we have, $S^r(\lim_{n \to \infty} x_{n+1}) = T^r(\lim_{n \to \infty} x_n) = 1$. Thus, $\lim_{n \to \infty} x_{n+1} = \lim_{n \to \infty} x_n = 2r$, which is the zero of $f^r x = x - 2r$

4.4 Exercises

> **Exercise D.1 1**
>
> (a) Take $s = 1$ in Theorem D.1, and state the Corollary arising from it
>
> (b) Take $Y = X$ and $S = id$, the identity map on X in Theorem D.1, and state the Corollary arising from it
>
> (c) Deduce that the corresponding result for the higher-order Mann iterative procedure obtained from Theorem D.1 is given by (b)

> **Exercise D.2 1**
>
> Chatterjea contraction was given in [Chatterjea SK: Fixed point theorems. C. R. Acad. Bulgare Sci. 1972, 25: 727-730]. Using techniques of this chapter, introduce a concept of Jungck-Chatterjea contractions and prove r-stability for higher-order Jungck-Mann iteration scheme in the setting of b-metric space using the higher-order version of the concept of Jungck-Chatterjea contractions.

> **Exercise D.3 1**
>
> Recall from [Ampadu, Clement (2015):Generalization of Higher-Order Contraction Mapping Theorem. Unpublished] that a map $S : X \mapsto X$ is called a higher-order Banach contraction if it satisfies: $d(S^r x, S^r y) \leq M \lambda^r d(x, y)$ for all $x, y \in X$ and any $r \in \mathbb{N}$, where $\lambda \in [0, 1)$ and $M \geq 1$ is the bound given by Proposition 4.1 contained in [Ezearn Fixed Point Theory and Applications (2015) 2015:88]. If we replace the identity with a continuous map, $T : X \mapsto X$ (say), then we obtain $d(S^r x, S^r y) \leq M^\star \lambda^r d(Tx, Ty)$, where $M^\star$ is a certain modification on M, and we call such a contraction a higher-order Jungck-contraction. A stability result for Jungck-contractions was given in b-metric space by Theorem 4.1 [S.L Singh and Bhagwati Prasad, Some coincidence theorems and stability of iterative procedures, Computers and Mathematics with Applications 55 (2008) 2512–2520] using the Jungck iterative procedure. Taking inspiration from this work, obtain a stability result for the higher-order Jungck contraction in b-metric space using the so-called higher-order Jungck iterative procedure given by Definition D.7. In particular, with respect to this chapter, what is the higher order version of Theorem 4.1 [S.L Singh and Bhagwati Prasad, Some coincidence theorems and stability of iterative procedures, Computers and Mathematics with Applications 55 (2008) 2512–2520] ?

4.5 References

(1) R. Kannan. Some results on fixed points II. American Mathematical Monthly, 76 (1969) 405-408

(2) S. L. Singh, C. Bhatnagar and S. N. Mishra, "Stability of Jungck-type iterative procedures," International Journal of Mathematics and Mathematical Sciences, vol. 2005, pp. 3035-3043, 2005

(3) W. R. Mann, "Mean value methods in iteration," Proceedings of the American Mathematical Society, vol. 4, pp. 506-510, 1953

(4) A. M. Harder and T. L. Hicks, A stable iteration procedure for nonexpansive mappings, Math. Japonica, vol. 33, pp. 687-692, 1988

(5) A.M. Harder and T. L. Hicks, Stability results for fixed point iteration procedures, Math. Japonica, vol. 33, pp. 693-706, 1988

(6) S. Czerwik, "Nonlinear set-valued contraction mappings in b-metric spaces," Atti Del Seminario Matematico E Fisico Universita Di Modena, vol. 46, pp. 263-276, 1998

(7) G. Jungck, "Commuting mappings and fixed points," Am. Math. Mon., vol. 83, pp. 261–263, 1976

(8) Chatterjea SK: Fixed point theorems. C. R. Acad. Bulgare Sci. 1972, 25: 727-730

(9) Ampadu, Clement (2015):Generalization of Higher-Order Contraction Mapping Theorem. Unpublished

(10) Ezearn Fixed Point Theory and Applications (2015) 2015:88

(11) S.L Singh and Bhagwati Prasad, Some coincidence theorems and stability of iterative procedures, Computers and Mathematics with Applications 55 (2008) 2512–2520

Chapter 5

Stability of Common r-Fixed Point Procedure for Higher-Order Generalizations of Banach, Kannan, and Chatterjea Contractions

5.1 Brief Summary

> **Abstract E.1 1**
>
> We investigate r-stability for some generalizations of Banach, Kannan, and Chatterjea contractions involving four self-maps

5.2 Preliminaries

> **Definition E.1 1**
>
> Let T be a self-map of a metric space X, the set of r-fixed points of T for any $r \in \mathbb{N}$ will be denoted by $\{x \in X : T^r x = x\}$

> **Definition E.2 1**
>
> Let the sequence $\{x_n\}$ converging to the r-fixed point of T in X be generated by the iterative procedure $f(T^r, x_n)$ for any $r \in \mathbb{N}$, where f is some function. We will say $f(T^r, x_n)$ is r-T-stable or r-stable with respect to T iff a sequence $\{y_n\}$ in X, approximately close to $\{x_n\}$, converges to the same r-fixed point in X

CHAPTER 5. STABILITY OF COMMON R-FIXED POINT PROCEDURE FOR HIGHER-ORDER GENERALIZATIONS OF BANACH, KANNAN, AND CHATTERJEA CONTRACTIONS

Remark E.3 1

The formal definition of 1-stability was given in [A.M. Harder and T.I. Hicks, A stable iteration procedure for non-expansive mappings, Math. Japonica, 33, No.5(1988),687-692; A.M. Harder and T.I. Hicks, Stability results for fixed point iteration procedures, Math. Japonica, 33, No.5(1988), 693-706], and since then many authors have studied several special cases of the general iterative procedure over many years, for examples see [V. Berinde, Iterative Approximation of Fixed Points, Editura Efemeride, Baia Mare, (2002); C.O. Imoru, M.O. Olatinwo, Some stability theorems for some iteration procedures, Acta Univ. Palacki. Olomuc., Fac. rer. nat., Mathematica 45(2006), 81-88; J.R. Jachymski, Common fixed point theorems for families of maps, Indian J. Pure Appl. Math. 25, No.9(1994), 925-937; J.R. Jachymski, An extension of A. Ostrowski's theorem on the round-off stability of iterations, Aequationes Math. 53, No.3(1997), 242-253; J.Matkowski and S.L. Singh, Round off stability of functional iterations on product spaces, Indian J. Math. 39, No.3(1997), 275-286; M.O. Osilike, Stability results for fixed point iteration procedures, J. Nigerian Math. Soc. 14/15(1995/96),17-29; B. E. Rhoades, Fixed point theorems and stability results for fixed point iteration procedures II Indian J. Pure Appl. Math. 24, No.11(1993), 691-703]

Definition E.4 1

Let $S, T : Y \mapsto X$, $T^r(Y) \subseteq S^r(Y)$, and z be a r-coincidence point of T and S, that is, $S^r z = T^r z = p$ (say) for any $r \in \mathbb{N}$. For any $x_0 \in Y$, let the sequence $\{S^r x_n\}$, generated by the iterative procedure $S^r x_{n+1} = f(T^r, x_n)$, $n = 0, 1, 2, \cdots$ and any $r \in \mathbb{N}$, converge to p. Let $\{S^r y_n\}$ in X be an arbitrary sequence, and set $\epsilon_n = d(S^r y_{n+1}, f(T^r, y_n))$, $n = 0, 1, 2, \cdots$ and any $r \in \mathbb{N}$. Then the iterative procedure $f(T^r, x_n)$ will be called r-(S,T)-stable or r-stable with respect to (S,T) if and only if $\lim_{n \to \infty} \epsilon_n = 0$ implies $\lim_{n \to \infty} S^r y_n = p$

Remark E.5 1

1-stability of the procedure in the above definition was developed in [S.L. Singh, C. Bhatnagar, S.N. Mishra, Stability of Jungck-type iteration procedures, Int. J. Math. Math. Sci. 19(2005) 3035-3043]. By putting $Y = X$ and $f(T, x_n) = T x_n$, yields the Jungck iteration procedure (or J-iteration) in the literature, namely, $S x_{n+1} = T x_n$, $n = 0, 1, 2, \cdots$, and it gives the Picard iterative procedure in the literature when S is taken to be the identity map on X

Definition E.6 1

Let A, B, S, T be mappings of a metric space (X, d) into itself such that $A^r X \subseteq T^r X$ and $B^r X \subseteq S^r X$ for any $r \in \mathbb{N}$. Let $x_0 \in X$ be arbitrary and define a sequence $\{y_n\}$ in X such that $y_{2n} = A^r x_{2n} = T^r x_{2n+1}$ and $y_{2n+1} = S^r x_{2n+2} = B^r x_{2n+1}$ for all $n = 0, 1, 2, \cdots$ and any $r \in \mathbb{N}$

(a) The iterative procedure will be called numerically r-stable iff a sequence $\{z_n\}$ in X, approximately close to $\{y_n\}$, converges to the common r-fixed point $p \in X$ of A,B,S, and T

(b) Define $\epsilon_{2n} = d(B^r z_{2n+1}, S^r z_{2n})$ and $\epsilon_{2n+1} = d(A^r z_{2n}, T^r z_{2n+1})$ for all $n = 1, 2, \cdots$ and any $r \in \mathbb{N}$, then the iterative procedure will be called r-(A, B, S, T)-stable or r-stable with respect to (A, B, S, T) if and only if $\lim_{n \to \infty} \epsilon_n = 0 \Leftrightarrow \lim_{n \to \infty} S^r z_{2n} = \lim_{n \to \infty} T^r z_{2n+1} = p$

CHAPTER 5. STABILITY OF COMMON R-FIXED POINT PROCEDURE FOR HIGHER-ORDER GENERALIZATIONS OF BANACH, KANNAN, AND CHATTERJEA CONTRACTIONS

> **Remark E.7 1**
>
> The iterative sequence in the above definition when $r = 1$ has been investigated by many authors, for examples, see [R.P. Pant, Common fixed points of four mappings, Bull. Cal. Math Soc. 90(1998), 281-286; H.K. Pathak, M.S. Khan, S.M. Kang, Fixed and coincidence points for contraction and Parametrically nonexpansive mappings, Math. Sci. Res. J., 8, No. 1 (2004), 27-35]. They have used it to establish existence of unique common fixed points for some classes of contractions in both metric and Banach spaces.

Recall from [Ampadu, Clement (2015):Generalization of Higher-Order Contraction Mapping Theorem. Unpublished] that a map $T : X \mapsto X$, is called a higher-order Banach contraction if it satisfies $d(T^r x, T^r y) \leq M\lambda^r d(x,y)$ for all $x, y \in X$ and any $r \in \mathbb{N}$, where $\lambda \in [0,1)$ and $M \geq 1$ is the bound given by Proposition 4.1 contained in [Ezearn Fixed Point Theory and Applications (2015) 2015:88], and (X,d) is a metric space. If we put $A^r x := T^r x$, $B^r y := T^r y$, $Sx := x$, and $Ty := y$, then we obtain, $d(A^r x, B^r y) \leq M^\star \lambda^r d(Sx, Ty)$, where $M^\star$ is a modification on M. For this generalization of the Banach contraction, we obtain a stability result.

Recall from [Ampadu, Clement (2016):Stability of the Higher-Order Jungck-Kannan Mapping Operator in b-Metric Space. Unpublished] that given self-maps S, T of a metric space X, we say S is higher-order Jungck-Kannan contraction if it satisfies $d(S^r x, S^r y) \leq Z\beta^r[d(Tx, Sx) + d(Ty, Sy)]$ for all $x, y \in X$ and any $r \in \mathbb{N}$, where $Z \geq 1$ is given by Proposition 1.15[Ampadu, Clement (2016):Stability of the Higher-Order Jungck-Kannan Mapping Operator in b-Metric Space. Unpublished] and $\beta \in [0, \frac{1}{2})$. If we put $A^r x := S^r x$, $B^r y := S^r y$, $Sx := Tx$, $Ax := Sx$, $Ty := Ty$, and $By := Sy$, then, we obtain $d(A^r x, B^r y) \leq Z^\star \beta^r[d(Sx, Ax) + d(Ty, By)]$, where $Z^\star$ is a modification on Z. For this generalization of the Kannan contraction, we also obtain a stability result.

> **Definition E.8 1**
>
> Let (X,d) be a metric space and let the self-maps $A, B, S, T : X \to X$ satisfy $d(A^r x, B^r y) \leq \sum_{q=0}^{r-1} c_q[d(S^{q+1}x, B^{q+1}y) + d(T^{q+1}y, A^{q+1}x)]$ for all x, y in X and $r \in \mathbb{N}$. If $0 \leq c_q < \frac{1}{2}$ for all $0 \leq q \leq r-1$, then, we will say $A, B : X \to X$ form an rth order contraction with respect to $S, T : X \mapsto X$

> **Remark E.9 1**
>
> If $A = B$, $S = T = id$ (the identity map on X), and $r = 1$ in the previous definition, then $A = B : X \mapsto X$ is a Chatterjea contraction [Chatterjea SK: Fixed point theorems. C. R. Acad. Bulgare Sci. 1972, 25: 727-730], therefore the above definition is a generalization of Chatterjea contraction

> **Proposition E.10 1**
>
> Let (X,d) be a metric space, and let $A, B : X \mapsto X$ be an rth-order contraction with respect to $S, T : X \mapsto X$. For every pair $x \neq y \in X$, define
>
> $$J := J(x,y) = \max_{0 \leq v \leq r-1} \zeta^{-v} \frac{d(A^v x, B^v y)}{d(Sx, By) + d(Ty, Ax)}$$
>
> then
>
> $$J = \max_{n \in 0 \cup \mathbb{N}} \zeta^{-n} \frac{d(A^n x, B^n y)}{d(Sx, By) + d(Ty, Ax)}$$
>
> where $\zeta \in [0, \frac{1}{2})$

Now we have the following alternate characterization of Definition E.8

> **Definition E.11 1**
>
> Let (X, d) be a metric space. We will say $A, B : X \mapsto X$ is an rth-order contraction with respect to $S, T : X \mapsto X$, if $d(A^r x, B^r y) \leq J\zeta^r[d(Sx, By) + d(Ty, Ax)]$ for all $x, y \in X$ and any $r \in \mathbb{N}$, where $J \geq 1$ is given by the previous proposition and $\zeta \in [0, \frac{1}{2})$

For the generalization of the Chatterjea contraction, given by the above definition, we also obtain a stability result

5.3 Main Results

> **Theorem E.1 1**
>
> Let A, B, S, T be mappings of a metric space (X, d) into itself such that $A^r(X) \subseteq T^r(X)$ and $B^r(X) \subseteq S^r(X)$, and $S^r X$ or $T^r X$ is a complete subspace of X, for any $r \in \mathbb{N}$. Let p be a common r-fixed point of A, B, S, T, and the sequence $\{y_n\}$ generated by $y_{2n} = A^r x_{2n} = T^r x_{2n+1}$ and $y_{2n+1} = S^r x_{2n+2} = B^r x_{2n+1}$ for all $n = 0, 1, 2, \cdots$, $r \in \mathbb{N}$, and $x_0 \in X$ converge to p. Let $\{z_n\}$ be in X and define $\epsilon_{2n} = d(B^r z_{2n+1}, S^r z_{2n})$ and $\epsilon_{2n+1} = d(A^r z_{2n}, T^r z_{2n+1})$, $n = 0, 1, \cdots$, and any $r \in \mathbb{N}$. If A, B, S, T satisfy $d(A^r x, B^r y) \leq M^\star \lambda^r d(Sx, Ty)$, for all $x, y \in X$, where $M^\star$ is a modification on M, $\lambda \in [0, 1)$, and $M \geq 1$ is the bound given by Proposition 4.1 contained in [Ezearn Fixed Point Theory and Applications (2015) 2015:88], then, $\lim_{n \to \infty} \epsilon_n = 0 \Leftrightarrow \lim_{n \to \infty} S^r z_{2n} = \lim_{n \to \infty} T^r z_{2n+1} = p$

Proof of Theorem E.1 1

Suppose $\lim_{n \to \infty} \epsilon_n = 0$, $n = 0, 1, 2, \cdots$. Now observe that

$$\begin{aligned}
d(p, T^r z_{2n+1}) &\leq d(T^r z_{2n+1}, A^r z_{2n}) + d(A^r z_{2n}, B^r x_{2n+1}) + d(B^r x_{2n+1}, p) \\
&\leq \epsilon_{2n+1} + M^\star \lambda^r d(S^r z_{2n}, T^r x_{2n+1}) + d(B^r x_{2n+1}, p) \\
&\leq \epsilon_{2n+1} + M^\star \lambda^r [d(S^r z_{2n}, B^r z_{2n+1}) + d(B^r z_{2n+1}, T^r x_{2n+1})] + d(B^r x_{2n+1}, p) \\
&= \epsilon_{2n+1} + M^\star \lambda^r \epsilon_{2n} + M^\star \lambda^r d(B^r z_{2n+1}, A^r x_{2n}) + d(B^r x_{2n+1}, p) \\
&\leq \epsilon_{2n+1} + M^\star \lambda^r \epsilon_{2n} + (M^\star \lambda^r)^2 d(S^r z_{2n}, T^r x_{2n+1}) + d(B^r x_{2n+1}, p)
\end{aligned}$$

Since as $n \to \infty$, we have the following, $\epsilon_n \to 0$, $S^r x_{2n} \to p$, and $B^r x_{2n+1} \to p$. If we take limits in the above inequality, and since $1 - (M^\star \lambda^r)^2 > 0$, we deduce that $T^r z_{2n+1} \to p$ as $n \to \infty$. Since $\{A^r z_{2n}\} \subseteq \{T^r z_{2n+1}\}$, we also have $\lim_{n \to \infty} A^r z_{2n} = p$. Now

$$\begin{aligned}
d(p, S^r z_{2n}) &\leq d(S^r z_{2n}, B^r z_{2n+1}) + d(B^r z_{2n+1}, A^r z_{2n}) + d(p, A^r z_{2n}) \\
&\leq \epsilon_{2n} + M^\star \lambda^r d(S^r z_{2n}, T^r z_{2n+1}) + d(p, A^r z_{2n})
\end{aligned}$$

Since as $n \to \infty$, we have $\epsilon_{2n} \to 0$, $A^r z_{2n} \to 0$ and $T^r z_{2n+1} \to p$. If we take limits in the above inequality, and since $1 - M^\star \lambda^r > 0$, we deduce that $\lim_{n \to \infty} S^r z_{2n} = p$. Therefore, we have $\lim_{n \to \infty} \epsilon_n = 0$ implies $\lim_{n \to \infty} S^r z_{2n} = \lim_{n \to \infty} T^r z_{2n+1} = p$. On the other hand if $\lim_{n \to \infty} S^r z_n = \lim_{n \to \infty} T^r z_n = p$, then since $\{A^r z_{2n}\} \subseteq \{T^r z_{2n+1}\}$ and $\{B^r z_{2n+1}\} \subseteq \{S^r z_{2n}\}$, we also have $\lim_{n \to \infty} A^r z_{2n} = \lim_{n \to \infty} B^r z_{2n+1} = p$. Thus, $\lim_{n \to \infty} \epsilon_n = 0$

Corollary E.2 1

Let A, B, S be mappings of a metric space (X, d) into itself such that $A^r X \cup B^r X \subseteq S^r X$, and $S^r X$ is a complete subspace of X for any $r \in \mathbb{N}$. Let p be a common r-fixed point of A, B, S and let the sequence $\{y_n\}$ in X generated by $x_0 \in X$ and $y_{2n} = A^r x_{2n} = S^r x_{2n+1}$ and $y_{2n+1} = S^r x_{2n+2} = B^r x_{2n+1}$, $n = 0, 1, \cdots$, and any $r \in \mathbb{N}$ converge to p. Let $\{z_n\}$ be in X, and define $\epsilon_{2n} = d(B^r z_{2n+1}, S^r z_{2n})$ and $\epsilon_{2n+1} = d(A^r z_{2n}, S^r z_{2n+1})$, $n = 0, 1, \cdots$, and any $r \in \mathbb{N}$. If A, B, S satisfy $d(A^r x, B^r y) \leq M^{\star\star} \lambda^r d(Sx, Sy)$, for all $x, y \in X$, where $M^{\star\star}$ is a modification on $M^\star$, $M^\star$ is a certain modification on M, $\lambda \in [0, 1)$, and $M \geq 1$ is the bound given by Proposition 4.1 contained in [Ezearn Fixed Point Theory and Applications (2015) 2015:88], then, $\lim_{n \to \infty} \epsilon_n = 0 \Leftrightarrow \lim_{n \to \infty} S^r z_n = p$

Proof of Corollary E.2 1

It is necessary only to take $S^r = T^r$ for any $r \in \mathbb{N}$ in the previous theorem

Theorem E.3 1

Let A, B, S, T be mappings of a metric space (X, d) into itself such that for any $r \in \mathbb{N}$, $A^r X \subseteq T^r X$ and $B^r X \subseteq S^r X$, and $S^r Y$ or $T^r Y$ is a complete subspace of X. Suppose p is a common r-fixed point of A, B, S, T and the sequence $\{y_n\}$ in X generated by $y_{2n} = A^r x_{2n} = T^r x_{2n+1}$ and $y_{2n+1} = S^r x_{2n+2} = B^r x_{2n+1}$ for all $n = 0, 1, 2, \cdots, r \in \mathbb{N}$, and $x_0 \in X$ converge to p. Let $\{z_n\}$ be in X and define $\epsilon_{2n} = d(B^r z_{2n+1}, S^r z_{2n})$ and $\epsilon_{2n+1} = d(A^r z_{2n}, T^r z_{2n+1})$, $n = 0, 1, \cdots$, and any $r \in \mathbb{N}$. If A, B, S, T satisfy at least one of Definition E.11 and $d(A^r x, B^r y) \leq Z^\star \beta^r [d(Sx, Ax) + d(Ty, By)]$, where $Z^\star$ is a modification on Z, $Z \geq 1$ is given by Proposition 1.15[Ampadu, Clement (2016):Stability of the Higher-Order Jungck-Kannan Mapping Operator in b-Metric Space. Unpublished] and $\beta \in [0, \frac{1}{2})$, then $\lim_{n \to \infty} \epsilon_n = 0 \Leftrightarrow \lim_{n \to \infty} S^r z_{2n} = \lim_{n \to \infty} T^r z_{2n+1} = p$

Proof of Theorem E.3 1

Assume $d(A^r x, B^r y) \leq Z^\star \beta^r [d(Sx, Ax) + d(Ty, By)]$ holds for all $x, y \in X$. Suppose $\lim_{n \to \infty} \epsilon_n = 0$. Now observe that,

$$d(A^r z_{2n}, S^r z_{2n}) \leq d(A^r z_{2n}, B^r z_{2n+1}) + d(B^r z_{2n+1}, S^r z_{2n})$$
$$\leq Z^\star \beta^r [d(S^r z_{2n}, A^r z_{2n}) + d(T^r z_{2n+1}, B^r z_{2n+1})] + \epsilon_{2n}$$

From the above one deduces that

$$(1 - Z^\star \beta^r) d(A^r z_{2n}, S^r z_{2n}) \leq \epsilon_{2n} + Z^\star \beta^r [d(T^r z_{2n+1}, A^r z_{2n}) + d(A^r z_{2n}, B^r z_{2n+1})]$$
$$= \epsilon_{2n} + Z^\star \beta^r \epsilon_{2n+1} + Z^\star \beta^r d(A^r z_{2n}, B^r z_{2n+1})$$
$$\leq \epsilon_{2n} + Z^\star \beta^r \epsilon_{2n+1} + Z^\star \beta^r [d(A^r z_{2n}, S^r z_{2n}) + d(S^r z_{2n}, B^r z_{2n+1})]$$

Since $1 - 2Z^\star \beta^r > 0$, from the above one deduces that

$$d(A^r z_{2n}, S^r z_{2n}) \leq \frac{1 + Z^\star \beta^r}{1 - 2Z^\star \beta^r} \epsilon_{2n} + \frac{Z^\star \beta^r}{1 - 2Z^\star \beta^r} \epsilon_{2n+1}$$

Taking limits in above, we deduce that, $\lim_{n \to \infty} A^r z_{2n} = \lim_{n \to \infty} S^r z_{2n}$. If we take $x = z_{2n}$ and $y = x_{2n+1}$ in $d(A^r x, B^r y) \leq Z^\star \beta^r [d(Sx, Ax) + d(Ty, By)]$ and take limits, we deduce that, $\lim_{n \to \infty} A^r z_{2n} = p$. By the inclusions, $A^r X \subseteq T^r X$ and $B^r X \subseteq S^r X$, we have $\lim_{n \to \infty} S^r z_{2n} = \lim_{n \to \infty} A^r z_{2n} = \lim_{n \to \infty} T^r z_{2n+1} = p$. Conversely, let $\lim_{n \to \infty} S^r z_{2n} = \lim_{n \to \infty} T^r z_{2n+1} = p$, by the inclusions, $A^r X \subseteq T^r X$ and $B^r X \subseteq S^r X$, we also have $\lim_{n \to \infty} A^r z_{2n} = \lim_{n \to \infty} B^r z_{2n+1} = p$. Consequently, $\lim_{n \to \infty} \epsilon_{2n} = \lim_{n \to \infty} \epsilon_{2n+1} = 0$. On the other hand, if Definition E.11 holds, then, we deduce that $d(A^r z_{2n}, S^r z_{2n}) \leq (1 + J\zeta^r)\epsilon_{2n} + J\zeta^r \epsilon_{2n+1}$, and the rest of proof uses same technique as above.

We can summarize Theorem E.1 and Theorem E.3 in the following way

Corollary E.4 1

Let A, B, S, T be mappings of a metric space (X, d) into itself such that $A^r X \subseteq T^r X$ and $B^r X \subseteq S^r X$, and $S^r X$ or $T^r X$ is a complete subspace of X. Suppose p is a common r-fixed point of A, B, S, T and the sequence $\{y_n\}$ generated by $y_{2n} = A^r x_{2n} = T^r x_{2n+1}$ and $y_{2n+1} = S^r x_{2n+2} = B^r x_{2n+1}$ for all $n = 0, 1, 2, \cdots$, $r \in \mathbb{N}$, and $x_0 \in X$ converge to p. Let $\{z_n\}$ be in X and define $\epsilon_{2n} = d(B^r z_{2n+1}, S^r z_{2n})$ and $\epsilon_{2n+1} = d(A^r z_{2n}, T^r z_{2n+1})$, $n = 0, 1, \cdots$, and any $r \in \mathbb{N}$. If A, B, S, T satisfy $d(A^r x, B^r y) \leq \max\{M^\star \lambda^r d(Sx, Ty), Z^\star \beta^r [d(Sx, Ax) + d(Ty, By)], J\zeta^r [d(Sx, By) + d(Ty, Ax)]\}$, for all $x, y \in X$, where, $\lambda \in [0, 1)$, $\beta, \zeta \in [0, \frac{1}{2})$, $M^\star$ is a modification on M, and $M \geq 1$ is the bound given by Proposition 4.1 contained in [Ezearn Fixed Point Theory and Applications (2015) 2015:88], $Z^\star$ is a modification on Z, $Z \geq 1$ is given by Proposition 1.15[Ampadu, Clement (2016):Stability of the Higher-Order Jungck-Kannan Mapping Operator in b-Metric Space. Unpublished], and $J \geq 1$ is given by Proposition E.10, then, $\lim_{n \to \infty} \epsilon_n = 0 \Leftrightarrow \lim_{n \to \infty} S^r z_{2n} = \lim_{n \to \infty} T^r z_{2n+1} = p$

Remark E.5 1

If $A = B, S = T = id$, where "id" is the identity mapping of X in the contractive definition of the previous Corollary, then we get the higher-order version of the Zamfirescu mapping, and hence a r-stability result. More information about the Zamfirescu mapping can be found in [V. Berinde, Iterative Approximation of Fixed Points, Editura Efemeride, Baia Mare, (2002), B. E. Rhoades, A Comparison of various definitions of contractive mappings, Trans. Amer. Math. Soc., 226, (1977), 257-290, T. Zamfirescu, Fixed point theorems in metric spaces, Arch. Math., 23(1972) 292-298]

5.4 Exercises

Exercise E.1 1

The multi-step iteration scheme was introduced in [B. E. Rhoades and S. M. Soltuz, "The equivalence between Mann-Ishikawa iterations and multistep iteration," Nonlinear Anal., vol. 58, pp. 219–228, 2004], and here we introduce a multi-step scheme to approximate the common r-fixed point of the higher-order Jungck contraction mapping [Exercise D.3] in Banach space: Let $x_0 \in Y$, the higher-order Jungck-multi-step scheme associated with the sequence $\{S^r x_n\}$ for any $r \in \mathbb{N}$ will be defined as, $S^r x_{n+1} = (1 - \alpha_n) S^r x_n + \alpha_n T^r y_n^1$; $S^r y_n^i = (1 - \beta_n^i) S^r x_n + \beta_n^i T^r y_n^{i+1}$, $i = 1, 2, \cdots, k-2$; $S^r y_n^{k-1} = (1 - \beta_n^{k-1}) S^r x_n + \beta_n^{k-1} T^r x_n$, $k \geq 2$, where $\{\alpha_n\}_{n=0}^{\infty}$ and $\{\beta_n^i\}_{n=0}^{\infty}$, $i = 1, 2, \cdots, k-1$ are real sequences in $[0, 1)$ such that $\sum_{n=0}^{\infty} \alpha_n = \infty$

Prove the following: Let X be a Banach space and $S, T : Y \mapsto X$ for an arbitrary set Y such that $\|T^r x - T^r y\| \leq M^{\star\star} \lambda^r \|Sx - Sy\|$ holds, where $M^{\star\star}$ is a certain modification on $M^\star$[Exercise D.3], and $T^r Y \subseteq S^r Y$. Assume that S and T have a r-coincidence point z such that $T^r z = S^r z = p$. For any $x_0 \in Y$, the higher-order Jungck-multistep scheme associated with $\{S^r x_n\}$ converges to p. Further if $Y = X$ and S, T r-commute at p[$S^r p = T^r p$, then $S^r T^r p = T^r S^r p$], then p is the unique common r-fixed point of S, T[$S^r p = T^r p = p$].

HINT: J.O. Olaleru and H.Akewe, On Multistep Iterative Scheme for Approximating the Common Fixed Points of Contractive-Like Operators, International Journal of Mathematics and Mathematical Sciences, Volume 2010, Article ID 530964, 11 pages

Exercise E.2 1

The Jungck-Noor iteration scheme was considered in [Olatinwo M.O., A generalization of some convergence results using the Jungck-Noor three step iteration process in arbitrary Banach space, Fasc. Math., 40(2008), 37-43] and here we introduce a higher-order Jungck-Noor scheme to approximate the common r-fixed point of the higher-order Jungck contraction mapping [Exercise D.3] in Banach space: Let $x_0 \in Y$. The higher-order Jungck-Noor iteration scheme associated with the sequence $\{S^r x_n\}_{n=1}^{\infty}$ for any $r \in \mathbb{N}$ will be defined as, $S^r x_{n+1} = (1-\alpha_n)S^r x_n + \alpha_n T^r y_n$; $S^r y_n = (1-\beta_n)S^r x_n + \beta_n T^r z_n$; $S^r z_n = (1-\gamma_n)S^r x_n + \gamma_n T^r x_n$, where $\{\alpha_n\}_{n=0}^{\infty}$, $\{\beta_n\}_{n=0}^{\infty}$, and $\{\gamma_n\}_{n=0}^{\infty}$ are real sequences in $[0,1)$ such that $\sum_{n=0}^{\infty} \alpha_n = \infty$

Prove the following: Let X be a Banach space, $S, T : X \mapsto X$ be such that $\|T^r x - T^r y\| \leq M^{\star\star} \lambda^r \|Sx - Sy\|$, where $M^{\star\star}$ is a certain modification on $M^\star$[Exercise D.3], and $T^r X \subseteq S^r X$. Assume S,T are r-weakly compatible [for some $p \in X$, $S^r p = T^r p$, then $S^r T^r p = T^r S^r p$]. For any $x_0 \in X$, the higher-order Jungck-Noor iteration scheme associated with $\{S^r x_n\}_{n=1}^{\infty}$ converges to the unique common r-fixed point of S, T.

HINT: J.O. Olaleru and H.Akewe, THE CONVERGENCE OF JUNGCK-TYPE ITERATIVE SCHEMES FOR GENERALIZED CONTRACTIVE-LIKE OPERATORS, Fasciculi Mathematici, Nr 40 (2010),87-98

Exercise E.3 1

Here we introduce a higher-order version of the modified multistep scheme [H.Akewe, MODIFIED MULTISTEP ITERATION FOR APPROXIMATING A GENERAL CLASS OF FUNCTIONS IN LOCALLY CONVEX SPACES, Acta Math. Univ. Comenianae Vol. LXXXIII, 1 (2014), pp. 39-45] and use it to approximate the r-fixed point for a class of functions satisfying a higher-order Banach contractive type condition [Theorem B.1] in locally convex spaces: Let X be a metrisable topological space and C be a nonempty subset of X, and T be a self-map of C. Let $x_0 \in C$, the higher-order modified multi-step iteration scheme associated with $\{x_n\}$ will be defined as $x_{n+1} = (1-\alpha_n)y_n^1 + \alpha_n T^r y_n^1$; $y_n^i = (1-\beta_n^i)y_n^{i+1} + \beta_n^i T^r y_n^{i+1}$; $y_n^{p-1} = (1-\beta_n^{p-1})x_n + \beta_n^{p-1} T^r x_n, i=1,2,\cdots,p-1, p \geq 2$, where $\{\alpha_n\}$, $\{\beta_n^i\}$, $i=1,2,\cdots,p-1$ are real sequences in $[0,1]$ such that $\sum_{n=0}^{\infty} \alpha_n = \infty$

Prove the following: Let (X, f_c) be a complete metrisable locally convex space, K a closed convex subset of X, and $T : K \mapsto K$ be an operator with r-fixed point p satisfying the condition $f_c(p - T^r y) \leq M^\star \lambda^r f_c(p - y)$, for each $y \in K$, where $M^\star$ is a certain modification on the bound given by Proposition 4.1[Ezearn Fixed Point Theory and Applications (2015) 2015:88] and $\lambda \in [0,1)$. For $x_0 \in K$, let the higher-order modified multi-step iteration scheme associated with $\{x_n\}$ converge to p. Then the higher-order modified multi-step iteration scheme converges strongly to p

5.5 References

(1) A.M. Harder and T.I. Hicks, A stable iteration procedure for nonexpansive mappings, Math. Japonica, 33, No.5(1988),687-692

(2) A.M. Harder and T.I. Hicks, Stability results for fixed point iteration procedures, Math. Japonica, 33, No.5(1988), 693-706

(3) V. Berinde, Iterative Approximation of Fixed Points, Editura Efemeride, Baia Mare, (2002)

(4) C.O. Imoru, M.O. Olatinwo, Some stability theorems for some iteration procedures, Acta Univ. Palacki. Olomuc., Fac. rer. nat., Mathematica 45(2006),81-88

(5) J.R. Jachymski, Common fixed point theorems for families of maps, Indian J. Pure Appl. Math. 25, No.9(1994), 925-937

(6) J.R. Jachymski, An extension of A. Ostrowski's theorem on the round-off stability of iterations, Aequationes Math. 53, No.3(1997), 242- 253

(7) J.Matkowski and S.L. Singh, Round off stability of functional iterations on product spaces, Indian J. Math. 39, No.3(1997), 275-286

(8) M.O. Osilike, Stability results for fixed point iteration procedures, J. Nigerian Math. Soc. 14/15(1995/96),17-29

(9) B. E. Rhoades, Fixed point theorems and stability results for fixed point iteration procedures II Indian J. Pure Appl. Math. 24, No 11(1993), 691-703

(10) S.L. Singh, C. Bhatnagar, S.N. Mishra, Stability of Jungck-type iteration procedures, Int. J. Math. Math. Sci. 19(2005) 3035-3043

(11) R.P. Pant, Common fixed points of four mappings, Bull. Cal. Math Soc. 90(1998), 281-286

(12) H.K. Pathak, M.S. Khan, S.M. Kang, Fixed and coincidence points for contraction and Parametrically nonexpansive mappings, Math. Sci. Res. J., 8, No. 1 (2004), 27-35

(13) Ampadu, Clement (2015):Generalization of Higher-Order Contraction Mapping Theorem. Unpublished

(14) Ezearn Fixed Point Theory and Applications (2015) 2015:88

(15) Ampadu, Clement (2016):Stability of the Higher-Order Jungck-Kannan Mapping Operator in b-Metric Space. Unpublished

(16) Chatterjea SK: Fixed point theorems. C. R. Acad. Bulgare Sci. 1972, 25: 727-730

(17) B. E. Rhoades, A Comparison of various definitions of contractive mappings, Trans. Amer. Math. Soc., 226, (1977), 257-290

(18) T. Zamfirescu, Fixed point theorems in metric spaces, Arch. Math., 23(1972) 292-298

(19) B. E. Rhoades and S. M. Soltuz, "The equivalence between Mann-Ishikawa iterations and multistep iteration," Nonlinear Anal., vol. 58, pp. 219–228, 2004

(20) J.O. Olaleru and H.Akewe, On Multistep Iterative Scheme for Approximating the Common Fixed Points of Contractive-Like Operators, International Journal of Mathematics and Mathematical Sciences, Volume 2010, Article ID 530964, 11 pages

(21) Olatinwo M.O., A generalization of some convergence results using the Jungck-Noor three step iteration process in arbitrary Banach space, Fasc. Math., 40(2008), 37-43

(22) J.O. Olaleru and H.Akewe, THE CONVERGENCE OF JUNGCK-TYPE ITERATIVE SCHEMES FOR GENERALIZED CONTRACTIVE-LIKE OPERATORS, Fasciculi Mathematici, Nr 40 (2010),87-98

(23) H.Akewe, MODIFIED MULTISTEP ITERATION FOR APPROXIMATING A GENERAL CLASS OF FUNCTIONS IN LOCALLY CONVEX SPACES, Acta Math. Univ. Comenianae Vol. LXXXIII, 1 (2014), pp. 39-45

www.ingramcontent.com/pod-product-compliance
Lightning Source LLC
Chambersburg PA
CBHW051109180526
45172CB00002B/842